ISO 14001 Environmental Systems Handbook

ISO 14001 Environmental Systems Handbook

Second edition

Ken Whitelaw

ELSEVIER
BUTTERWORTH
HEINEMANN

AMSTERDAM • BOSTON • HEIDELBERG • LONDON • NEW YORK • OXFORD
PARIS • SAN DIEGO • SAN FRANCISCO • SINGAPORE • SYDNEY • TOKYO

Elsevier Butterworth-Heinemann
Linacre House, Jordan Hill, Oxford OX2 8DP
30 Corporate Drive, Burlington, MA 01803

First published 1997
Reprinted 2000
Transferred to digital printing 2003
Second edition 2004

British Library Cataloguing in Publication Data
A catalogue record for this book is available from the British Library

Library of Congress Cataloguing in Publication Data
A catalogue record for this book is available from the Library of Congress

ISBN 0 7506 4843 0

For information on all Elsevier Butterworth-Heinemann publications visit our website at http://books.elsevier.com

Printed and bound in Great Britain

Contents

Preface

The first edition of *ISO 14001 Implementation Handbook* was published in November 1997. Many changes have since occurred which are now reflected in this second edition, namely:

1. ISO 14001:1996 has been through a lengthy revision process.
2. Other Standards that were referenced in the first edition have themselves been revised and republished i.e. (ISO 9001:2000), and various health and safety standards have been consolidated into OHSAS 18001:1999.
3. At the time of publication of the first edition there was approximately 2000 ISO 14001 certificates world-wide. This has grown considerably, not only in the more well-developed countries, but also in emerging economies.
4. The uptake of the Standard by 'service industries' has continued unabated, and the book now widens its scope to address this rather than focusing on manufacturing industry.

5 Integration of management standards is becoming the norm. Organizations have come to realize the reduction in costs to the business and the improvements in efficiency, obtained from aligning their various management systems into one blueprint for the business. This has been helped in many ways by the Standards, as referenced above, which are now aligned with ISO 14001 in their structure.

6 The concept of added value (although its definition is imprecise) is now firmly entrenched within the certification industry. External auditors are expected not only to assess compliance with ISO 14001, but to also hand on best practice in the form of observations and opportunity for improvements to the implementing organization – the client. Clients rightly expect their systems to come under scrutiny but whatever the outcome, compliance or non-compliance, expect the auditor to give constructive criticism and suggest positive solutions. This is amplified in later chapters.

7 There is a higher awareness within industry of environmental issues and thus much of the 'lay history' of environmental concerns that was in the first edition has been condensed. More emphasis has been placed upon practical implementation of an environmental management system, hereafter abbreviated to EMS.

8 Since 1997, when the first edition was published, advances in information gathering via the Internet has expanded exponentially and this is reflected in this second edition by including website and email addresses in Appendix III.

Additionally, in the first edition of this book, the subject of legislation was not treated in depth, the author believing that there were other means by which organizations could address legislative issues. Due to requests from many parties – readers, potential readers and colleagues, the treatment of legislation has been expanded. Having expanded this issue, the reader is reminded that legislation changes relentlessly and only the foundation of building legislative compliance into an EMS is offered. Updating must be the responsibility of the implementing organization.

Ken Whitelaw

Introduction

History of environmental management

The environmental management system 'industry' has its experts – trainers, consultants, auditors and certification bodies – and they generally take for granted that the clients they deal with have at least as much understanding about EMS development as they have. During the training, consultancy or auditing process, there will be gaps in the knowledge of the client and errors in the implementation processes. That is why such experts are used by clients and called in to assist.

But do the experts devote as much time as they should do to explain the reasons how and why ISO 14001 evolved and is in the present format. Do clients have a whole host of misconceptions about why ISO 14001 is structured as it is? Do they believe (cynically) it was written to be only accessible and understood by environmental professionals and provide employment to the above mentioned disciplines? There is some anecdotal evidence to suggest that this cynicism is not wholly misplaced and that not enough has been said, written, discussed and mooted to allay some of these suspicions and illustrate the map of the evolution of environmental management; to show that there are historical triggers and drivers, fuelled by the interest of a host of stakeholders all seeking to reach the same goal – reconciling the demands of a modern technological society with the available resources of the planet. Such an understanding can only improve any organization's ability to manage for the environment.

This introduction will demonstrate that a combination of national and international forces at work, plus the legislative measures taken by successive governments, the rise of stakeholder power and green pressure groups, as well as the trigger of environmental disasters, have all played their part in the development of environmental management. The following are some of these forces:

National and international forces of change

Industrial Revolution – old and new environmental issues

We do not need to look too far back in modern history to find a point at which major impacts on the Earth's resources began and the balance of nature was disturbed. Certainly this occurred in the nineteenth century in Europe with mechanization – the Industrial Revolution. This was a period when inventiveness and innovation was at its height, and the resultant mechanization of manufacturing processes began to have negative impacts upon the environment. Prior to this period, any negative environmental impacts tended to be localized due to lack of mechanization. Immense changes to society began to occur and, consequently, vast amounts of non-renewable resources were consumed to support this industrialized society with little thought as to the longer-term effects on the health of the population or the quality of the environment.

Successive governments brought in legislative measures to control the worst excesses of manufacturing pollution (mainly due to health problems) and as the processes became more diversified and sophisticated in the twentieth century, ever more legislation became necessary to control these diverse industrial activities. Concurrent with this build-up of legislation, the powers of the policing authorities grew – especially with regard to imposing greater financial penalties. It also became clear that the impact – the magnitude – was also rising with disasters not restricted to national boundaries. In the twenty-first century many of the Industrial Revolution's pollution problems have receded especially in highly regulated, industrialized countries. However, new environmental impacts have taken their place – such as global warming and destruction of the ozone layer.

Towards the end of the twentieth century, the time became right for society to reflect that managing for the environment was necessary. In the business world specifically, at board meetings, management of the environment was an agenda item. In many ways, environmental issues began to be treated in the same way as commercial business decisions i.e. reactive management based upon the risk to future profits.

Increasing legislative penalties

A plethora of legislation now exists and is being continuously added to or amended. Indeed, keeping up to date with such legislation is difficult. Principles such as 'making the polluter pay' were established in the courts, making the organization that created the pollution responsible for all the clean-up costs, including any consequential damages.

Increasing financial penalties

These are now the norm including:

1. *Ethical investment*: The concept of only lending money to an organization that has demonstrated that it is environmentally responsible or, conversely, not lending money to an organization that has shown disregard for the environment.
2. *Insurance risk*: Insurance companies are looking very hard at the everyday risks they are routinely underwriting. An organization without a clear environmental policy may well be refused insurance cover or be required to pay higher premiums.

The rise of the power and influence of the stakeholder

At one time the term 'stakeholder' tended to mean an investor – a shareholder – of a business. When considering the environment, this definition of the stakeholder now refers to a whole spectrum of society who have an interest in the well-being of a business – and do not want such status damaged by poor environmental performance. Such a poor environmental performance may damage the image of the business (making it less attractive to investors), depress share prices, reduce market share, produce less profits and, perhaps, necessitate staff redundancies or close down a factory rendering employees redundant.

This spectrum of stakeholders includes:

- The parent company
- The board of directors
- The shareholders
- The employees
- The insurance company
- The regulatory body
- Customers
- Environmental pressure groups
- Suppliers
- The local community
- Competitors
- The general public

The general public are becoming more aware of 'green' issues; children learn about environmental concerns at school. The media focus on environmental events. The general public can demonstrate their environmental stance by exercising their power as purchasers. Such behaviour is based on the information available to them – usually via the media whose coverage may veer towards the sensational rather than a true account of events.

Any organization should perform a simple stakeholder analysis – it can be illuminating – to demonstrate just whom they are responsible to, and perhaps to ask what these stakeholders require of them!

Market encouragement measures

Market encouragement measures are measures designed to promote 'green' purchasing decisions on the part of the consumer. The main intention is to provide sufficient information for decision-makers to be properly informed about the environmental performance of goods and services. ECO labelling and establishment of an EMS are just two examples of this approach.

Disasters – as environmental triggers

Environmental events or disasters no matter where in the world, are given immediate and full coverage by the media. Such crises act as 'milestones' in the development of environmental awareness at the general public level and the enforcement of tougher legislation at the commercial level. The human tragedy of such disasters is the most emotive of course and this is closely followed by considerations of the environmental issues. Such disasters from the twentieth century are remembered by:

- Place names: Bhopal, India (fugitive emission of methyl isocyanate gas)
- Oil tankers: Exxon Valdez (widespread marine pollution)
- Legislation: 'The Seveso Directive' (fugitive emission of dioxins)

Disasters such as these are extreme examples of environmental catastrophes and certainly prompted major enquiries and initiated new legislation aimed at prevention and repetition of the disaster.

Sustainability

The general public is becoming increasingly aware of the issue of sustainability. The very science and technology that has been developed to give us the life style that we enjoy now, especially in the Western industrial world, has left us in a position to choose what environmental legacy to hand down to our children and grandchildren.

ISO 14001 – evolution from previous environmental standards

One of the consequences of the above issues and concerns was a large number of requests to standard-writing bodies to produce a standard for managing the environmental impacts of an organization. As a result, the British Standards Institute (BSI), a world-respected standards body, in conjunction with many other committees and interested parties, developed and produced BS 7750:1992, the world's first environmental standard. Other similar national standards were in existence in various countries and world-wide demand for accredited certification to an international standard began to grow.

ISO (the International Standards Organization) established a new technical committee to develop international standards in environmental management. The need for the standard to be applicable to manufacturing and services industries was heeded as was fulfilling the needs of all sizes of businesses. The need to avoid trade barriers as well as the different approaches to legal requirements and their enforcement throughout the world demanded a generic approach.

Thus ISO 14001 was first published in 1996 with a swift uptake by organizations world-wide.

All ISO standards are periodically reviewed, typically between 3 and 5 years. ISO 14001 was reviewed informally by ISO at the appropriate time but it was felt that there was insufficient hard experience of world-wide use to justify a major revision to ISO 14001.

Thus the revised ISO 14001 has no major changes. However, the brief of the committee responsible was to:

1 Ensure compatibility with ISO 9001:2000 which was a major review of ISO 9000:1994

2 Improve the clarity of the text i.e. clarity of intent especially for translation purposes

Reasons for seeking ISO 14001 certification

The reasons why organizations implement ISO 14001 are generally given as:

- To gain or retain market share via a green corporate image
- To attract more ethical investment
- To reduce insurance risks
- To reduce prosecution risks
- To reduce costs

The reasons may not necessarily be in this order of importance. However, the fact is that cost savings tend to be low on the list of responses. Image and potential loss of business are cited the most frequently. Cost savings tend to be overlooked, yet this is an area where implementing organizations can have major benefits.

SMEs

The above is relevant to all companies but it is recognized that as the larger organizations achieve ISO 14001, there is then pressure exerted downwards to the supplier chain inevitably towards the SMEs. SMEs react quicker to downturns in economies and may well place environmental management lower in the priorities when survival of the business is at stake. Nevertheless, research over the years has shown a higher willingness to implement environmental management from SMEs. This book is therefore aimed towards those managers of medium-sized companies which have the potential to impact upon the environment in a measurable sense. This can include the smaller organizations, i.e. the SMEs.

Acknowledgements

As an environmental auditor employed by an international certification body, SGS United Kingdom Ltd, I am grateful to the many implementing organizations that I have worked with around the world, as an auditor, trainer and advisor. I am particularly grateful for the valuable contributions from the organizations in Chapter 5.

Extracts taken directly from ISO 14001 are reproduced with kind permission of the British Standards Institute, and complete editions of this and other related standards can be obtained from them. Their address is included in Appendix III.

KW
May 2004

Chapter 1

Concepts and the 'spirit' of ISO 14001

Introduction

Before describing the steps an organization needs to take to implement the Standard itself, some of the underlying concepts are considered so that an understanding may be reached as to why the clauses of the Standard are written as they are. An understanding of the intention of the Standard – the 'spirit' – is also considered. For, unless the requirements of the Standard are understood at an early stage, the resultant EMS may have weak foundations. Such a system will not give the performance improvements intended – thus wasting the resources of the implementing organization.

The structure and the purpose of the clauses and sub-clauses are addressed in straightforward language and, where possible, simple, illustrative examples are given. This chapter sets out the framework whereas Chapter 2 details all the steps necessary for practical implementation of ISO 14001. The first parts of this chapter explore the concepts behind what any EMS should set out to achieve. Later, attention is focused on how such concepts are refined for ISO 14001 environmental management systems: the reasons for the clauses; why they are phrased in the way that they are; and what they require of an organization in practical terms.

Concepts of environmental control

On a personal level

We all have an impact on the environment by the mere act of living from day-to-day. An EMS, in its simplest form, asks us to control our activities so that any environmental impacts are minimized. This broad and simplistic approach has its merits. However, such a loose, unstructured approach may lead us to improve in the wrong direction or, indeed, may leave us without any clear direction at all. On a personal level, it is tempting to control and minimize those impacts we feel we can tackle easily. Perhaps our attitude towards environmental issues is influenced by a topical environmental event, and therefore, we can be influenced to act without thoroughly understanding some of the more complex issues.

Thus, as individuals, we may focus on, and minimize, environmental impacts which are trivial in nature compared with other impacts (which are far more significant and require more considered thought processes). As an example, we may commit ourselves to a futile exercise without attacking the root causes of pollution or the use of non-renewable energy sources. We may, in our working environment, always: re-use paper (writing on both sides); re-use paper clips; recycle plastic drinks cups. Such measures require only a little thought, and very little personal physical effort. Yet we may use our car to drive to the office in a city – contributing to air pollution, traffic congestion and so on – when alternative transport could be used (for example, the humble bicycle or public transport). However, this latter option for environmental control requires much forethought (such as planning the journey time around bus timetables). There is some personal inconvenience and physical effort in this choice as well as some loss of freedom and flexibility. This is not to say that re-use of paper clips should be discarded as an environmentally responsible option but that we must be aware of its environmental significance compared to the other, more significant, environmental impacts.

At a business level

Moving from individual actions to corporate actions, and using the analogy above, unless a structured approach is taken the organization may focus on what it believes to be its environmental impacts, a belief based upon 'gut feel' and ease of implementation. In reality, this does not address real issues but promotes a 'green' feel-good factor or perceived enhancement of image – both internal and external to the organization – which is not justified. For example, a company engaged in the extraction of raw materials by mining may have an environmental objective to save energy. By implementing a 'save energy by switching off lights' campaign in its site offices it may feel it has achieved 'green' status and may proudly boast of such an environment-friendly approach.

There will be some energy saved by administration personnel switching off lights and heating when they are not being used for long periods. However, such savings in energy are trivial compared to the massive impact that the mining industry has on the environment: the visible impact of the site and surrounding land; the associated increased noise levels from the operation of such a site; the high use of energy both in extraction technology and transport activities; the use of chemicals in the purification process; and of course, the use of non-renewable resources (the raw material that is being mined). Unless the mining company considers the relative scale and significance of environmental impacts, then by claiming to be 'green' it has really missed the whole point of environmental control and impact minimization.

Thus, this concept of *significance* is fundamental and must be at the heart of any environmental management system. An organization must move away from this 'gut feel' approach to a structured system that demands as a minimum from the organization, an understanding of the concepts behind and strong linkages between:

- Identifying all environmental aspects of the organization's activities
- Using a logical, objective (rather than subjective) methodology to rank such aspects into order of significant impact upon the environment
- Focusing the management system to seek to improve upon and minimize such significant environmental impacts

It should be noted that the criteria used for attributing significance to environmental impacts should be clearly defined. The process of evaluating each aspect against the criteria should be readily apparent. The 'significance value' of an impact can be a numeric one, an alphanumeric one or a significance rating resulting from an informed decision-making process undertaken by a team, or even one person. This methodology of rating of significance is very important – it must be robust and withstand scrutiny, and be reproducible during the life of the EMS. Rating is examined in greater detail in Chapter 2.

For example, for one company the most significant environmental impact could be the sending of mixed waste to a landfill site – a fairly common environmental aspect shared by most manufacturing organizations. The organization must then decide on an objective to aim for in reduction of such waste. This objective could be to reduce progressively waste sent to landfill by 3% per annum. Individual targets to support this objective could be set to: progressively segregate waste; recycle a certain percentage; and, perhaps, sell off a certain percentage of segregated waste (for example, brown cardboard). All these measures would reduce the number of skips of waste sent to the landfill site.

It should be noted the percentage savings, for example, should have an attainable figure based upon what is practicable in the situation and what other similar industries are achieving i.e. benchmarking. The organization should use easy-to-measure data to support this waste minimization objective and this is further examined, developed and discussed in Chapter 2. Remembering that any EMS is seeking to place controls upon its environmental impacts, then it is only common sense to have a plan for monitoring and measurement of controlling activities. Such a plan should readily show any deviations from the targets during a review, so that if a problem does occur, then the appropriate remedial or corrective actions can easily be taken.

This is environmental control. ISO 14001 provides the framework to allow such controls to be exercised in a structured and controlled way. By documenting such a system, personnel operating it have a framework to: work around; hang ideas onto; follow what is documented; record what was done; and learn from any mistakes that were made.

The spirit of ISO 14001

In the simplest of terms, and condensing the whole concept of ISO 14001 into one sentence, we can say that fundamentally the Standard requires an organization to:

> *Control and reduce its impact on the environment.*

In simple terms, the Standard requires an organization to state how it goes about controlling and reducing its impact on the environment: doing in practice what it has stated in its environmental policy; recording what has occurred; and learning from experience.

What obligation does this impose upon an organization? ISO 14001 requires an organization to control its impacts on the environment. All aspects of business activity cause changes in the environment to a greater or lesser extent. Organizations deplete energy sources and raw materials and generate products and waste materials. These changes are referred to as *environmental impacts*. ISO 14001 defines an environmental impact as:

> *Any change to the environment, whether adverse or beneficial, wholly or partially resulting from an organization's activities, products or services.*

Identifying and assessing the significance of environmental impacts is a critical stage in an organization's preparatory stages for ISO 14001. Thus the organization needs to understand that by operating its processes, by manufacturing its products or supplying its services, it is depleting natural resources and using non-renewable energy sources. At the same time it is also producing by-products in the form of waste materials.

This should not, however, promote guilty feelings within the organization! The Standard does not require organizations to feel guilty and apologetic. There is no hidden agenda to close the business down. The Standard requires management, by forethought and action, to use less scarce resources by better planning, use recycled materials and perhaps operate the process differently. An element of the controls required by the Standard will be dictated by the demands of legislation. Thus, to keep within the law, the organization will wish to ensure that all regulatory and legislative requirements concerning its environmental performance are satisfied. Increasingly, however, organizations are seeking to go beyond those legal requirements in order to ensure that their environmental integrity (of activities, products and services) meets the expectations of the stakeholders. So, in effect, compliance with the law is mandated by the legal authorities. Controlling environmental impacts is also mandated – not by the legal authorities but by the stakeholders – as there is an inherent requirement, from the above discussion, to improve or minimize environmental impacts.

During the period of planning the implementation, some organizations have wondered how the ISO 14001 system will operate at the point in time when all the environmental objectives of the organization have been fulfilled and where, perhaps, further improvements would be subject to the law of diminishing returns. What does the organization do next? Will ISO 14001 certification be lost? Does the organization attempt to improve in environmentally trivial areas, performing a meaningless paperwork exercise merely to generate evidence that the system is still alive, in order to retain certification?

The reality is that once the initial significant environmental impacts have been controlled and minimized, the other hitherto less significant impacts become more significant and a new cycle of improvement begins. Thus the cycle is never-ending and there is continuous improvement of the organization's environmental performance.

Two illustrations from history demonstrate (with hindsight) that our knowledge of environmental issues is usually flawed and that we, as individuals and organizations, acted in an environmentally responsible way based upon the knowledge available to us at that time:

- The use of CFCs (ozone-depleting chemicals) was not thought to be an environmental issue. We now know that it has a highly significant global environmental impact with possible long-term damage to our quality of life on Earth.

- Similarly, the widespread use of asbestos was at one time not thought to be an environmental issue nor a safety hazard.

Thus, the rules can change. New knowledge comes to light and new, tougher legislation will always be around the corner. Therefore, this status of 'zero or trivial significant environmental impacts' will never occur.

It is also tempting for a cynic of environmental management to compare two similar organizations manufacturing the same products. Although they manufacture the same products, one of them is noted for having a higher impact upon the environment than the other:

- Producing more waste to landfill

- Using more energy due to older plant

- Has more breaches of legislation – violations of discharge consents, for example
- Is visually offensive due to old, badly sited buildings
- Has more smell and noise nuisance

The cynic will ask how can both organizations achieve ISO 14001 if one appears to be not as environmentally responsible as the other? The answer is that they are both equally environmentally responsible if they are certified to ISO 14001. They are both equally committed to environmental improvement but the starting point for this environmental improvement is different for each of them.

They will both have the same potential environmental impacts but the landfill and energy-usage question may be due to better or worse technology – one organization may have access to capital from a parent company and will therefore perform better in these respects than their poorer competitor.

The environmental improvement objectives of the 'poorer' company may, in fact, be similar to their more affluent competitor but, for example, percentage improvement figures may be of a lower order. One longer-term objective could be to match the environmental performance prevalent within the organization's own industrial sector. This objective is very much dependent upon an organization's economic performance.

An EMS does not seek to be comparable – it proves only that each organization is seen to be committed to taking appropriate and practical steps to reduce their environmental impacts (within their individual capability and level of technology).

Providing that both organizations can demonstrate such commitment, the certification body will allow certification. This is the concept of the EMS: it is an improvement process, rather than a method for stating that, at any one point in time, one organization is performing better than another.

The clauses of the Standard evolve from this simple common sense investigation of an organization's activities with some additions and enhancements (for example, ensuring mechanisms are in place to make a company aware of new and impending legislation). As the reader will note from reading the Standard itself, it is not a long document and is written concisely, with only six main clauses. It is generic in style, as it is intended to be applicable to any manufacturing or service industry.

The following section looks at the Standard in more detail.

Clauses of ISO 14001

Having looked at the concepts of environmental system management and the intentions of ISO 14001, this section discusses, in broad terms, the six clauses of ISO 14001. It explains their intended purpose, prior to a fuller examination in Chapter 2.

ISO 14001 requires organizations to identify the environmental aspects of their activities, products or services and to evaluate the resulting impacts on the environment, so that objectives and targets can be set for controlling significant impacts and for improving environmental performance. ISO 14001 specifies the EMS requirements that an organization must meet in order to achieve certification by a third party – the certification body. (The Standard was specifically designed to be an auditable standard leading to independent certification.)

The requirements of ISO 14001 include:

- Development of an environmental policy
- Identification of environmental aspects and evaluation of associated environmental impact
- Establishment of relevant legal and regulatory requirements
- Development and maintenance of environmental objectives and targets
- Implementation of a documented system, including elements of training, operational controls and dealing with emergencies
- Monitoring and measurement of operational activities
- Environmental internal auditing

- Management reviews of the system to ensure its continuing effectiveness and suitability

The informative Annex A of ISO 14001 contains additional guidance on the use of the Standard. Annex B contains a matrix of the linkages and cross-references between ISO 14001 and ISO 9001 and will probably be of interest to those organizations who wish to combine these two separate management systems. This is addressed further in Chapter 4.

However, before venturing into the clauses, there is one area of implementation which appears to have been left out of the Standard – and can be considered to be clause 4.0 (if one wishes). This is the Preparatory Environmental Review (PER). A *preparatory environmental review* is an investigative exercise – a structured piece of detective work – which identifies all of the organization's environmental aspects and is addressed in more detail in Chapter 2. This initial step is not mandatory and cannot be audited during the assessment (see Chapter 3) and yet, if it is not performed, the whole environmental management system may not be soundly based. An organization may have a clear vision of where it would like to be in terms of future environmental performance. However, unless this 'snapshot' of current performance is undertaken – the PER – the organization may act in an unfocused manner and not achieve this goal.

So unless an organization knows where it is *now* with regard to its interaction with the environment, it may not be able to move in the correct direction (forward) in controlling and minimizing its environmental impacts. It is only after performing a preparatory environmental review that a meaningful environmental policy, with proper and relevant objectives and targets, can be set out.

The six main clauses of ISO 14001 are titled as follows:

4.1 General Requirements

4.2 Environmental Policy

4.3 Planning

4.4 Implementation and Operation

4.5 Checking and Corrective Actions

4.6 Management Review

These are now discussed with emphasis on the concepts of the clauses.

Clause 4.1: General requirements

Annex A.1 within the ISO 14001 specification, which is applicable to this clause of the Standard describes, at length, the intended purpose of the EMS (that is, an improvement in the environmental management system is intended to show an improvement in environmental performance).

The words used in the Standard are 'establish' and 'maintain'. There is no guidance as to the level or depth of establishment or what evidence is required to show maintenance. However, it can be taken to mean that there must be some objective evidence of the system being in place and reviewed and revised. Such review processes can take the form of monthly progress meetings, corrective actions from audits and of course audits themselves. Audits are evidence of reviewing; that is, asking the question: 'Are the planned activities of the organization occurring in practice?'

It would be very prudent for an implementing organization to offer evidence of – at the very least – one or two audits of a significant impact over a period of 2 to 3 months. Keeping documentary evidence of management reviews (even if not the fully formalized 'Management Review' that clause 4.6 of the Standard refers to) and having some evidence of awareness training for those staff who control those environmental aspects that have a highly significant environmental impact, would also be wise steps to take.

Clause 4.2: Environmental policy

The intention of this requirement of the Standard was that by making the organization's environmental policy available to the public, the organization was very clearly setting highly visible environmental objectives. By this, the organization demonstrated commitment and accountability which could be verified, examined, and even criticized, if it failed to deliver the promises made. Thus, the policy was intended to be the main 'driver' of the environmental management system and all other elements of the system would follow on naturally from it.

This clause includes the statement 'be available to the public'. In fact, the concept of public availability is open to many interpretations and in reality it is only the stakeholders who are interested in the content of the environmental policy (rather than the general public).

Annex A.2 of the Standard refers to the policy to be 'within the context of the environmental policy of any broader corporate body': so that the policy of the organization will not be in conflict with higher-level corporate strategy so that coherence of policy could be seen to flow throughout the wider organization. This was to ensure that no policies of higher-level management would prevent the policies of lower-level management being effectively carried out, for example, at the site level.

The environmental policy must also be reviewed by top management. This was to ensure that the ultimate responsibility for, and commitment to, an EMS belongs to the highest level of management within the organization.

Because the environmental policy documentation is so fundamental, the meaning of the words on it must be clear and very relevant to the organization's activities. This is explored in depth in Chapter 2.

Clause 4.3: Planning

Through its sub-clauses, clause 4.3 directs the implementing organization towards key areas in the planning process. Thus:

Clause 4.3.1: Environmental aspects

The intent behind this sub-clause was to ensure that an organization had the capability and mechanisms to identify continually any environmental aspects it had, and then to attach a level of significance to those aspects in a structured and logical way. It is to be noted that Annex A.3.1 within ISO 14001:2004 does state that there is no one method for identifying environmental aspects and Chapter 2 gives examples of some of these mechanisms.

Because the environmental behaviour of a supplier, or indeed a customer, could well turn out to be not of the same level of responsibility exercised by the implementing organization, such 'indirect' or remote activities may well be of far more significance than that of the 'direct' impacts of the organization itself. It therefore makes sense for an organization to include such 'indirect' environmental aspects within its system and, using the same methodology, attach a level of significance. This can be especially pertinent to service orientated organizations.

Clause 4.3.2: Legal and other environmental requirements

This sub-clause requirement was included in the Standard because it was recognized by the authors of ISO 14001 that an organization could fall down on its environmental performance if it did not possess sufficient knowledge of applicable environmental laws, or codes of practice, within its industry sector. These codes of practice are the 'other environmental requirements'.

First and foremost, an organization must comply with local and national legislation. By definition, legislation exists to control significant environmental impacts, otherwise the legislation would not have come into being as was discussed in the Introduction. Thus, because of this implied significance, compliance with legislative requirements is the baseline for certification to ISO 14001. Many industrial sectors have membership bodies (such as the CIA – the 'Chemical Industry Association') who issue codes of practice to their members. These are generally guidance notes to ensure 'best practices' are followed. They tend to emphasize health and safety, and increasingly, environmental issues. One can be sure that, if a code of practice exists, there must be sound reasons why it was written. An organization would be expected to comply with it or to demonstrate compelling reasons why such a code had been disregarded.

Clause 4.3.3: Objectives, targets and programmes

Although the organization may have an environmental policy (derived from consideration of its preparatory environmental review) and may have identified those aspects of its business which have a significant environmental impact, it needs to translate such findings into clear achievable objectives, measured by specific targets. In practical terms, each significant environmental impact should have an associated objective (and targets) set against it for the control of, and minimization of, that impact. This then is the intention and purpose of this particular sub-clause.

Further, although the overall aim of the EMS should be continual improvement, not all the objectives have to relate to immediate improvements in environmental performance. However, such objectives must ultimately demonstrate overall improvement in performance. For example, there could be cases where improvements are identified as an objective but this objective may only be realized several years in the future when, perhaps, either investment or technology currently being developed will be available.

So there may well be a range of objectives: some completed in the short term, others being reached only in the longer term.

The sub-clause goes on to say that programmes are to be established and maintained; the purpose being to ensure that the organization has allocated responsibilities and resources and set time-scales for ensuring that the activities described in the preceding sub-clauses will happen as planned. To allow such a programme to be monitored, it therefore makes sense for the organization to document and make visible, and available, such a plan or programme to all involved employees.

Clause 4.4: Implementation and operation

This, the longest section in the Standard, has no less than seven sub-clauses, and is written to enable an organization to operate an environmental management system to the requirements of the Standard on what might be referred to as an every-day basis.

These sub-clauses are:

Clause 4.4.1: Resources, roles, responsibility and authority
This sub-clause was included to ensure that personnel are assigned specific responsibilities for a part, or parts, of the EMS and have a very clear-cut reporting structure (with no ambiguities). For example, when monitoring emissions to atmosphere, it should be clear who actually performs this task, with contingency plans for responsibility if the named person is away ill or on vacation.

History shows us that when an event turns into a crisis, the cause is often due to the fact that no one individual takes ownership of a problem. Some crises are caused by mismanagement in the form of disregard for safety procedures, for example, but in the majority of incidents, the people concerned did not know they were responsible. Either there was no clarity of roles or interfaces of responsibility were blurred.

Therefore, with this in mind, the sub-clause also requires that top management appoint an individual to be the 'management representative', with specific ownership for the well-being of the EMS and co-ordination of all environmental activities.

Clause 4.4.2: Competence, training and awareness
This sub-clause is designed to enable an organization not only to identify training needs, as appropriate, but also to measure the success of that training. All individuals need some form of training to enable them to perform a new task. In the context of ISO 14001, awareness is the product (or end result) of any training given. This enhanced knowledge should enable

them to make more informed decisions when dealing with environmentally related issues.

This informed decision-making is referred to as competence again within the context of ISO 14001. The organization must find mechanisms to test individuals competence because, although the audience in a training session may all appear to be equally receptive to the information given, there may well be big differences in performance when the knowledge from the training is put into practice.

Guidance in the Annex (A.4.2) does indicate that levels of training, and competence, be related to the significance of the individuals to influence environmental impacts within the organization.

Clause 4.4.3: Communication

It was believed by the authors of the Standard that means of communication, both internally and externally, are extremely important and that, if not formally addressed, may have negative effects on the success of the EMS.

Internal communications could be carried out through team briefings at regular intervals. Such briefings would inform all staff of the progress of the implementation of ISO 14001. At a later date these briefings, or the use of dedicated noticeboards, internal emails or internal newsletters, could inform all staff on the progress of the environmental objectives.

Unless there is a procedure in place stating responsibilities and planned dates for the above, the communication system could well become *ad hoc* and not be very effective.

External communications in most organizations tend to focus on the methodology for dealing with environmental complaints from neighbours.

But external communications can also relate to how the organization will respond to enquiries from all stakeholders: requests for information etc. on the environmental impacts of the organization or its products for example. Such communication needs to be co-ordinated to ensure that a consistent message is sent to stakeholders and a documented procedure would ensure this.

The organization, should also decide, perhaps within this documented procedure, whether it does want to communicate information on its impacts, or what depth of data it is willing to supply. The organization may well decide to publish an annual environmental report for example, and a procedure should be in place to assign responsibilities and deadlines for publication.

Clearly, in the event of an environmental emergency, the organization also needs to formulate the process of communication with the regulatory bodies, the media and the stakeholders.

Clause 4.4.4: Documentation

For a system to be audited, there must be a minimum level of documentation (consistent with the requirements of the Standard) available to demonstrate that the system exists, i.e. has been established. This documention can be hard copy or in the form of electronically held data.

The sub-clause itself uses phrases such as 'describing the main elements' which indicates that a top-heavy documented system is not the aim of this sub-clause. There is encouragement from the accreditation bodies and certification bodies alike to ensure that an EMS is not too focused on documentation alone. A balance must be struck between failure to document essentials and a bureaucratic system that does not add any value or meaning to the system. For example, where ISO 14001 calls for a 'documented' procedure, then clearly there must be such a document to achieve compliance with the Standard. In many other parts of the Standard, documenting a reference to existing documents – training manuals, Quality Assurance Procedures, machinery/plant operating instructions, etc. – will suffice. Such procedures could be in text format, flow or process charts.

Additionally, smaller organizations should note that if the environmental impacts of the organization are complex, then this complexity will dictate the level of documentation required to control and minimize such impacts. Straightforward environmental impacts may only require a modest level of documentation – reinforced by evidence of high levels of environmental awareness and competence being demonstrated by operating personnel.

Annex A4.4 indicates that the documentation need only provide direction. For example, if an organization has an established quality assurance system, then it will already have appropriate procedures for identifying training needs and where evidence of training (in the form of records) is filed. Such a procedure may then be referenced in the EMS. There is no need to duplicate what may be a perfectly adequate procedure. Again, accreditation bodies and certification bodies encourage such system integration. Showing cross-references to other management systems is also encouraged (for example, to occupational health and safety management systems).

It would be prudent for a simple matrix to be drawn up listing all the clauses and sub-clause headings of ISO 14001, and against each item on the list, the corresponding procedure title of the organization. This would show in a clear and visual manner whether all the requirements of the Standard had been addressed.

Undoubtedly this makes life easier for the external auditor when ensuring that the client has addressed all parts of the Standard, but the primary purpose of such a matrix is to help the client be confident that all parts of the Standard have indeed been addressed. The requirement that all clauses are addressed is mandatory for successful certification.

Clause 4.4.5: Control of documents

The purpose of document control in any management system is to ensure that when, for example, an employee follows a procedure, that procedure is the most up-to-date one available, and that an out-of-date procedure cannot be followed accidentally. In an environmental management system, following an outdated procedure could lead to adverse environmental consequences, so some method must be in place to control documentation.

In the Annex to the Standard (A.4.5), the relevant clause encourages organizations not to have a complex document control system. This is also the stance taken by accreditation and certification bodies.

Clause 4.4.6: Operational control

The purpose of operational control is to ensure that those environmental aspects that are deemed to be significant (as identified earlier in clause 4.3, Planning) are controlled in such a way that the objectives and targets have a fair chance of being achieved. Thus operational controls will tend to be prescriptive, and depending on the nature of the process or operation they refer to, could be detailed work instructions or process flow diagrams or charts. Such controls generally address the day-to-day operations of the organization.

The Standard calls for documented procedures but they need only be appropriate to the nature, complexity and degree of significance of the function, activity or process that they address.

If indirect environmental aspects have been evaluated as significant, then controls should be in place. This can include both suppliers and customers or the organization.

Clause 4.4.7: Emergency preparedness and response

The intent behind this sub-clause is that an organization must have in place plans of how to react in an emergency situation. Waiting until an emergency occurs and then formulating a plan is plainly not a good idea. The emergency plans or procedures may not work in practice, and this failure may lead to an environmental incident.

Thus it makes sense to identify the potential for an emergency, identify the risks and put plans in place to prevent and mitigate the environmental impacts associated with such an emergency. Several options are open to organizations, ranging from, at the simplest level, listing competent personnel who can be contacted (with alternatives) in the event of an out-of-hours emergency situation, to predicting worst-case scenarios that might involve serious pollution and perhaps loss of life. Such plans should be periodically tested, in the absence of genuine emergencies, to verify that they will work in practice.

Clause 4.5: Checking and corrective action

Checking refers to verifying that planned actions and activities take place. Thus a robust internal audit system (Clause 4.5.5) could be the method for this verification, but other mechanisms could be used such as reviews of reports indicating failures or delays to action plans.

Corrective action within an organization is required when the above checks demonstrate failures to meet targets, with preventive measures put in place to prevent recurrence of the same failure.

The following sub-clauses, when implemented, are key to achieving the above.

Clause 4.5.1: Monitoring and measurement

Monitoring in the sense of ISO 14001 means that the organization should check, review, inspect and observe its planned activities to ensure that they are occurring as intended.

So the management programme or programmes for environmental improvement cannot be said to be achieving anything unless the starting point is known, the objectives and targets are defined, and progress in between start and finish is monitored. Unless there is such regular monitoring, an environmental objective may not be achieved. Furthermore, the organization may not recognize this as a problem, nor take the necessary corrective actions.

Measurement is required to show absolute amounts of waste being produced, or recycled; percentage improvements in energy reduction; readings of pH meters to ensure compliance with legislation and so on.

Additionally, any equipment used during the measurement process must be reliable so that personnel using such equipment are confident that the readings shown are accurate. Such confidence in measuring equipment can be obtained by a systematic programme of calibration.

Clause 4.5.2: Evaluation of compliance

This sub-clause requires the organization to periodically evaluate compliance with current legislation. This was within several clauses of ISO 14001:1996 but was felt by many of the contributors to the drafting of the revision, that it should be given more emphasis in the form of its own sub-clause. No guidance is given as to what periodically means, but in the event of rapidly changing circumstances, the frequency of evaluation should be adjusted accordingly. The evaluation itself can take several forms with periodic audits of operational controls that are being used to ensure compliance with a discharge consent being just one example.

Clause 4.5.3: Non-conformity, corrective and preventive actions

Non-conformances, for example a failure to meet targets, must be recognized and acted upon. The root cause should be investigated and controls put in place to make sure the non-conformances do not happen again. Although this is the overriding purpose of this sub-clause, care must be taken to ensure that the corrective actions that are taken by the organization are commensurate with the environmental impact encountered and that committing excess time and resources to problems of a low magnitude is avoided.

Clause 4.5.4: Records

The purpose of this clause is to ensure that the organization keeps records of its activities. For example, in the event of a dispute with a regulatory body, not having records to demonstrate compliance with discharge consents (in the form of independent monitoring and measurement data) could spell trouble for the organization. A potentially heavy fine may be reduced if objective evidence in the form of records is produced which demonstrates due diligence. It therefore makes sense for the organization to decide which records it needs to keep, and for how long, commensurate with the risks involved if they did not keep such records. Legislative requirements will dictate that some records are kept for minimum specified time periods.

Clause 4.5.5: Internal audit

Internal audits are now an established management tool in many businesses. The concept of self-policing is recognized as an improvement mechanism by organizations with any form of management system. Environmental management systems are no different and this sub-clause requires that such audits are carried out. The audits should be carried out by, not only checking compliance of the organization to the requirements of ISO 14001, but also checking compliance to the organization's own procedures. Procedures may need to be revised to reflect current operational practice, or individuals reminded to follow the procedures.

Accreditation bodies insist that third-party certification bodies must determine the amount of reliance that can be placed upon the organization's internal audit. Such internal audits should be carried out in much greater depth than the external assessment body could hope to achieve and, this is an area upon which the certification body places much emphasis. The completeness and effectiveness of internal audits are major factors in demonstrating to the certification body that the environmental management system is being well managed.

Clause 4.6: Management review

The purpose of this clause is to consider, in a structured and measured way, all of the preceding steps that have been taken by the organization, and to ask fundamental questions such as:

- Is the organization doing and achieving what has been stated in the environmental policy?
- Are objectives and targets that are set for environmental performance being achieved?
- If objectives and targets are not achieved, why not?
- Are appropriate corrective actions taking place?

These questions, and more, should be asked by top management. The ideal vehicle for such an inward-looking review is a formalized management review with an itemized agenda, minutes being taken, and a report being issued to all interested parties. Management commitment is usually exhibited if indeed such reviews are attended by the senior management – site directors and the like who are, in the Standard's terminology, 'top' management.

A guideline for the time interval between reviews is 3 to 6 months in the early stages of implementation followed by annual reviews once the system becomes more mature. In reality, the time intervals should be determined by events.

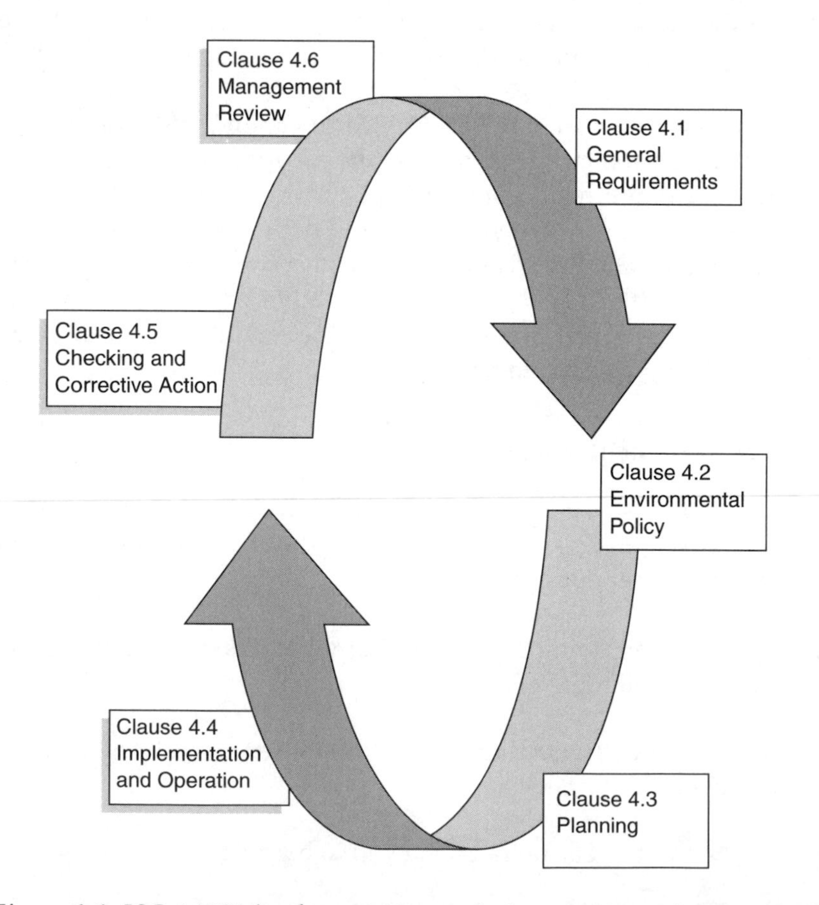

Figure 1.1 ISO 14001 implementation cycle for continuous improvement

For example, if it is found that the objectives and targets are being met, with very few exceptions, then the organization is well on its way to minimizing its environmental impacts and thus complying with the intended spirit of the Standard. Management reviews need only then take place annually. Conversely, many failures to achieve objectives demonstrate that the environmental management system has weaknesses and that more frequent management reviews should be held. Figure 1.1 illustrates the cyclic nature of continuous improvement required by the Standard, each 'cycle' culminating in the occurrence of a management review.

Summary

This chapter has looked at the concepts of environmental control. The point was made that being emotive about reducing environmental impacts may cause an organization to take a subjective approach to environmental management rather than one more focused on meaningful improvements in environmental performance by using the concept of significance.

This concept of 'significance' lies behind the wording of the Standard – its intention or 'spirit'. The actual words and phrases used are the product of the many international parties involved during the process of writing the Standard, who ensured that clarity was paramount. This clarity meant that the Standard could be audited against with some measure of consistency at an international level.

Chapter 2

Implementation of ISO 14001

Introduction

Chapter 1 introduced the fundamental concepts of ISO 14001 – why certain elements and requirements have been put into the Standard and why it is structured in the way that it is. This chapter builds up a more complete picture for the reader by addressing the detailed requirements of the Standard, clause by clause.

As before, illustrative examples are provided where appropriate. The intention of these examples is to stimulate thought as to the best way to interpret the particular clause of a generic standard for the reader's own organization.

First, a word is needed here on the concept of *commitment*. It is a word that is often misused. The effective implementation of the system can only be achieved if there is absolute commitment from top management. Such commitment includes the allocation of management time for the implementation phase as well as funding. Costs will be incurred during implementation and these must be budgeted for. Without a high input of such resources at the start of the implementation phase, the system will flounder and collapse.

This show of commitment is really no different to any other undertaking by the organization. If the junior personnel and staff see no commitment from the Managing Director to a new product or process, or development of a new market, they are hardly likely to be enthusiastic themselves.The resources of the organization will be stretched for some time, even if external consultants are used. Extra training will be required. Individual staff may have to perform tasks above and beyond their primary tasks. Therefore, management must ensure that this happens with some show of interest and enthusiasm and that the implementation process comes, temporarily at least, at number two or three on the business priority agenda: number one priority being maintaining a profitable business, of course.

How does an organization commence?

Commitment, as detailed above, is required from top management before the organization begins to implement an EMS. The Standard specifically refers to and includes requirements for 'resources, roles, responsibility and authority', and is seeking demonstration of a high level of management commitment (sub-clause 4.4.1 described later in this chapter).

Organizations are generally unsure of how long they can give this commitment for. The fact is, it is very difficult to give an optimum figure for the required length of time an organization will need for the implementation phase. However, based on the experience of both large and small companies, 12 months is too short a time period, whereas 3 years is too long. Eighteen months seems to be the 'norm'.

The reasons for this particular figure of 18 months are very much to do with people. Much of the success of any project's implementation is due to personnel having enthusiasm for the project's successful conclusion. People make it happen. However, time is required to change habits, to absorb the new culture and ways of working, to turn a new concept into part of everyday working practices.

In the early stages of a new EMS there is nothing very tangible for personnel to see. Therefore, if the time-scale is too protracted so that personnel cannot see an end or a goal being reached, then invariably enthusiasm wanes and the project tends to be put in the background of the hustle and bustle of every-day issues.

Cynicism may set in, personnel cannot see certification ever being gained, and the project fades away. Enthusiasm will never again be rekindled. So it appears that 18 months (and at the outside, 2 years) is the time-scale to aim for. The organization should 'switch on' the resources, if necessary, to reach this target.

It may be that an organization has one or more existing management systems. If so, it is possible – and perfectly acceptable from certification bodies' point of view – to use elements of such systems that already exist within the organization as part of the new environmental system (see Chapter 4 – Integration of Management Systems). Such existing systems could be quality assurance, occupational health and safety.

The preparatory environmental review will be discussed in detail now, prior to going into the details of the six main clauses of the Standard:

4.1 General Requirements

4.2 Environmental Policy

4.3 Planning

4.4 Implementation and Operation

4.5 Checking and Corrective Actions

4.6 Management Review

The preparatory environmental review (PER)

An organization considering the implementation of ISO 14001 should decide whether or not it needs to perform a Preparatory Environmental Review (PER).

This review is not mandated by the Standard, but in the Annex to the Standard (section A.3, Planning) it is suggested that an organization with no pre-existing environmental management system should establish its current position, with regard to the environment, by a review.

As stated in Chapter 1, the preparatory environmental review is not an auditable item on the external auditor's checklist but, by examining the review, the auditor will get a measure of the environmental competence of the organization and an indication of the level of understanding of environmental issues by the organization. In short, a level of confidence will be gained which can only assist in the smooth conduct of the assessment process (see Chapter 3).

Furthermore, if an organization has been operating an informal EMS for some time, it will in all probability have a sound knowledge of its environmental aspects, acquired by the experience of operating such a system. This will be true even if such an unfocused system does not meet the requirements of ISO 14001 itself. In essence, it has performed the equivalent of a preparatory environmental review.

Conversely, an organization new to the principles of environmental control or an EMS is well advised to take the opportunity to perform a structured review and to base their environmental management system upon the results of that review. If the decision is taken to move straight into implementation, there is a large body of evidence (from the certification bodies) that has shown that such systems are seriously flawed. The organization has had to rethink its strategy – with resultant setbacks in the certification time-scale.

If an organization decides to undertake this preparatory review, there are two options open:

1 Perform the preparatory environmental review using internally available resources.

2 Perform the preparatory environmental review using external consultants.

These options are considered below.

Internally available resources

Performing the preparatory environmental review using internally available resources has its merits in that the organization can use personnel experienced in the operations of its processes and, of some importance, costs can be somewhat better controlled.

Several options exist for this approach. One option is to send out questionnaires to each department head, requiring those individuals in charge to complete a series of questions, including, for example:

- What materials are used
- What quantities of materials are used
- How much energy is used
- The amount and type of waste streams
- Possible emergency situations
- Abnormal situations (frequency of start-ups and shut-downs, maintenance, breakdowns and incidents)
- Any history of 'out of the ordinary' incidents
- Any areas of training required

This will form a meaningful exercise by establishing the baseline to work from after analysis by management.

Possible flaws in this approach are that staff employed to perform this task (perhaps as an extension of their existing duties) may not have the necessary expertise to carry out a meaningful review. (Careful review of the answers in these completed questionnaires will indicate the level of training required in the personnel completing them!) Management tools such as 'brainstorming', although they are of value, will not give the same answers as hard data collection and some investigative detective work.

On occasions, organizations have made the mistake of basing their preparatory review on environmental projects that are currently up and running. The rationale is that if such projects are current, then they must be important, must focus on the significant environmental impacts of the organization and, therefore, must be a sound basis to start from. Unfortunately, although such projects may have been started with the best of intentions, they may be based upon previous initiatives (for example, a project that was topical at the time, or a project that looked easy to fulfil). The project may have been used to give credence to an individual or the organization during a marketing initiative at that time. Or again, it may have been a project which tied in with everyone's work schedules and was easy to manage, and with which everyone was comfortable because of the feeling that 'they were doing something for the environment'.

Such projects may well have been reducing environmental impacts and this is no reason to abandon them. Unfortunately, because of the haphazard nature and methodology of such projects they will, in all likelihood, not be focusing on significant impacts – fundamental to ISO 14001 philosophy.

It is essential if following this option that at least one senior manager in the organization has environmental expertise. If the expertise is not available within the organization, suitable existing staff might be trained by external consultants. This leads on to the second option.

External environmental consultants

Quite often the individual chosen to lead the environmental management team and implement ISO 14001 is the quality assurance manager. The reasons for this choice are not always valid. Those quality assurance managers who have taken on this task invariably perform well, but usually after a painful and steep learning curve. Just because the two Standards, ISO 14001 and ISO 9001 are now very much aligned, ISO 9001:2000 implementing experience may not bridge the shortfall in the required environmental knowledge required by the individual tasked with implementing ISO 14001.

There are of course commonalities between quality systems and environmental systems and a quality manager will certainly be comfortable in the areas of operating to documented management systems: the concept of objectives and targets and continuous improvement; the requirement for self-policing (auditing); the value of reviews; and a corrective and preventive action system to allow improvements to occur. But this does not necessarily equip the manager in question with the knowledge and skills in environmental issues required for ISO 14001. That said, the requirement for in-depth environmental knowledge from within the organization need not be onerous, and much of course depends upon the complexity of the organization's environmental aspects. A quality manager from a scientific background will probably be able to grasp environmental concepts more easily but, nevertheless, the organization must ask itself whether this is the correct choice of individual.

Knowledge of environmental legislation is an area where there could be a shortfall of information available from within an organization. In many countries, including the UK, there are organizations whose business is keeping up to date with legislation and providing such information for a fee to other organizations. This has immense value because it enables a smaller organization with limited resources to keep up to date with legislative issues.

However, to commence from a standing start into the complex world of environmental issues and perform a meaningful preparatory review is something not to be undertaken lightly. In such cases, the organization is well advised to use the services of an environmental consultancy and ask them to perform a preparatory review prior to ISO 14001 implementation.

Whichever route is taken, a well-executed preparatory environmental review will generate a 'specification' for the organization in the form of a report setting out very clearly what steps are required. This will form the foundation for deriving a meaningful environmental policy and developing a robust EMS, capable of demonstrating environmental performance improvement.

A typical format of a preparatory review is outlined below and an example of a checklist in Figure 2.1.

Steps to take – a checklist approach for a PER

The approach taken in the example of a PER below follows the informative advice supplied with the Standard in Annex A.3 and considers four key areas:

1. Legislative and regulatory requirements
2. Identification of significant environmental aspects
3. Examination of existing environmental practices and procedures
4. Assessment of previous incidents

Legislative and regulatory compliance

A fundamental requirement of ISO 14001 is that the organization complies with environmental law as a minimum standard. The review should identify

NOTES:	**SECTION A – Performed via site walkover**	**Review Findings**
Site Security	Perimeter fenced? Surveillance cameras? 24 hour security?	
Visual	Trees/shrubs/natural screening?	
Visual	Any adjacent footpaths/ rights of way?	
Visual	Adjacent neighbours/ area/ locality: Domestic? Industrial? Rural? Mixed?	
Visual Legal	Amenity areas? Parks? Fishing clubs? Ponds? Protected areas? SSSIs	
Legal Nuisance	Any noise detectable? Quantify: Loud/background? Type: Hi freq/Lo freq Occurrence: Intermittent/ Regular/constant	
Nuisance	Lights being used at night? Constant? Intermittent?	
Nuisance	Any odour detectable? Qualify - sweet? - pungent? - other?	
Visual/ Nuisance	Any dust/smoke detectable? Qualify? Light smoke? Dark smoke? Steam?	
Legal	Streams/Rivers/ Canals? Adjacent to site? Through the site? Site slopes towards/away?	

Depends upon neighbourhood and whether vandals are active

Natural screening reduces visual impact

Adjacent rights of way mean that members of the public could view the site and have a poor perception of the organization.

Industrial areas will not have as many nuisance-related environmental issues as rural or domestic

Intermittent noise can give rise to more complaints from neighbours!

Again, intermittent light 'nuisance' can give rise to complaints from neighbours

If the fall of the land of the site is towards a waterway, this should be highlighted as a potential environment impact

Figure 2.1 Preparatory environmental review checklist

Visual	Housekeeping generally? Visible to public? Obsolete machinery?	
Legal	Site drainage Drain covers marked?	Are surface water drains and foul sewer drains marked
Visual/ Legal/H & S	Storage areas: Drums? Silos? Tanks? Bunding around above? Underground tanks? Solvent stores? Any redundant tanks? Waste storage? Waste segregation?	Bunding/walls adequate and no damage? Adequate at 110% capacity? Redundant tanks – under or on the ground – have the potential to leak their residues over a period of time and are better removed
Legal	Contamination: Any evidence of small spillages? Any evidence of soil or hard standing Contamination? Likely contaminants? Electricity substations/transformers On site? Any visible signs of asbestos use On site?	Previous use of the site may have contaminated the land. See also section C of this checklist.
Nuisance/ Legal	Demolition on site Any demolition planned?	Dust and noise could be a nuisance to neighbours.
	SECTION B – Performed via internal walkover of offices, warehouses, manufacturing areas	**Review findings:**
Legal/ Visual/ Health and Safety	Storage areas: Raw Materials? Chemicals? Oils? Solvents? Fuels (diesel/petrol)?	Are COSHH sheets available at point of use?
Visual/ Legal	Evidence of any spillages?	Are spill kits available at point of use?

Figure 2.1 (Continued)

Legal	Segregation / Recycling wastes: Office: Warehouse: Factory:	Are paper, printer cartridges and batteries recycled
Legal/ Costs/ Environ-mental policy	Use of resources: Electricity?	Are there are areas poorly insulated against heat loss? Is there separate metering?
	Gas?	Where is gas used? In the process? Heating only?
	Water?	What processes is it used in? Is it recycled/reused?
	SECTION C – Performed by desktop study	**Review Findings**
	History of the site?	Previous site use could indicate contaminated land – may have impact upon future expansion plans for the organization or financial liabilities
	Future plans for the site?	Can have implications for the environmental policy and compliance with legislation
	Is the site shared with other organizations? Are there any shared facilities such as effluent treatment?	Planned demolition would need to take into account the site drainage system so that its integrity is not damaged
	Quantification of the above in sections A) and B) Waste Gas Electricity Water Raw materials	Use of the input / output diagram is of use here – Figure 2.2

Figure 2.1 (Continued)

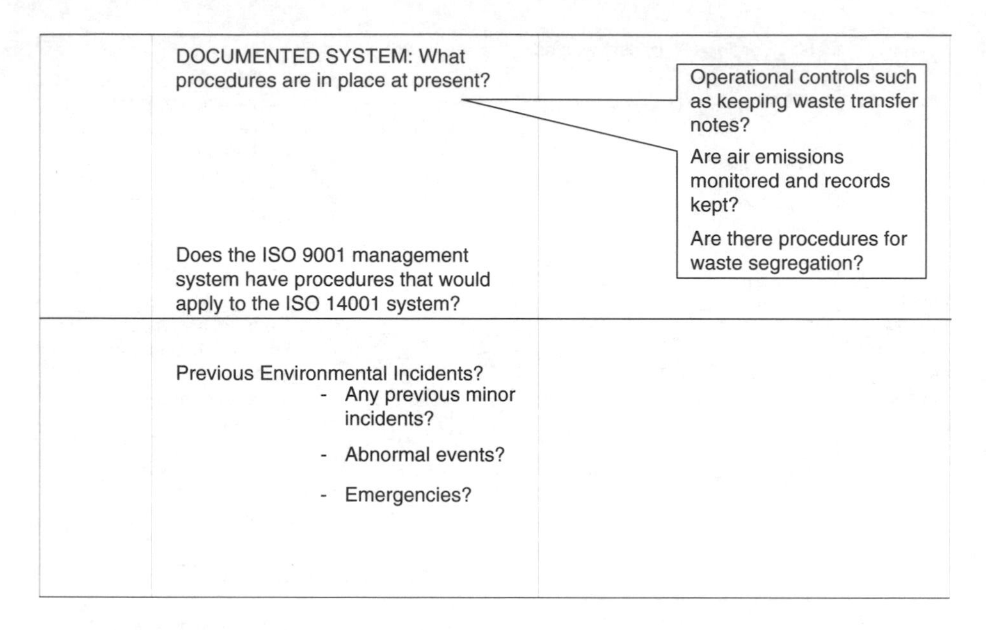

Figure 2.1 (Continued)

which areas of the organization are covered by which laws. Any areas where there are breaches of legislation should be set as priority action areas. Typical questions to ask are:

a) ***Is all existing legislation and other requirements being adhered to?***

Invariably there will be legislation relating to pollution or contamination of the three mediums air, land and water. Thus:

Air emissions:
If there is an authorized process, are the requirements for measuring, monitoring and recording being complied with?

Solid waste to land:
Is a waste management licence required for the site? If solid waste is taken away to landfill, do the operators of the landfill site have a licence? Does it cover the particular waste that is being removed? Does the carrier of the waste need a licence?

Water:
Is there any groundwater extracted on site? Is a licence required for this? Is any effluent discharged to streams, rivers, local authority

sewage systems? Is a licence to discharge required? If so, are the conditions being met?

Other requirements:
For example, does the site have any obligations to comply with any town and country planning consents or building regulations?

b) ***Is there any forthcoming legislation which may affect the business?***

For example, there may be legislation in the draft stage that, if enacted, could put the organization under heavy financial strain to comply. Therefore some investigative work is required.

c) ***Have there been previous incidents of breaches of legislation?***

Has the organization been prosecuted for any breach of environmental legislation? This may point to areas of weakness in the management system.

Evaluation of significant environmental aspects

An examination of an organization's environmental aspects is a key requirement of an EMS because unless all are identified, a potentially significant impact may be overlooked.Thus the use of a checklist plus an identification of the inputs and outputs of processes will be invaluable. Using an input–output type of diagram (see Figure 2.2) can assist in quantifying amounts of raw materials being used, energy consumption, levels of waste, etc. The input–output diagram could be performed for each office, each department or each process as appropriate.

Examination of existing environmental practices and procedures

There are many aspects to environmental management within an organization, especially management commitment. Such commitment is demonstrated in the environmental policy and documented procedures which are necessary to ensure such policies are known, understood and followed throughout the organization. The checklist (Figure 2.1 Section C) illustrates the scope of the documented management system that needs to be in place, with some typical records that need to be retained to demonstrate a minimum level of an EMS.

Assessment of previous incidents

The preparatory environmental review should also include an assessment of previous incidents – under both abnormal and emergency situations.

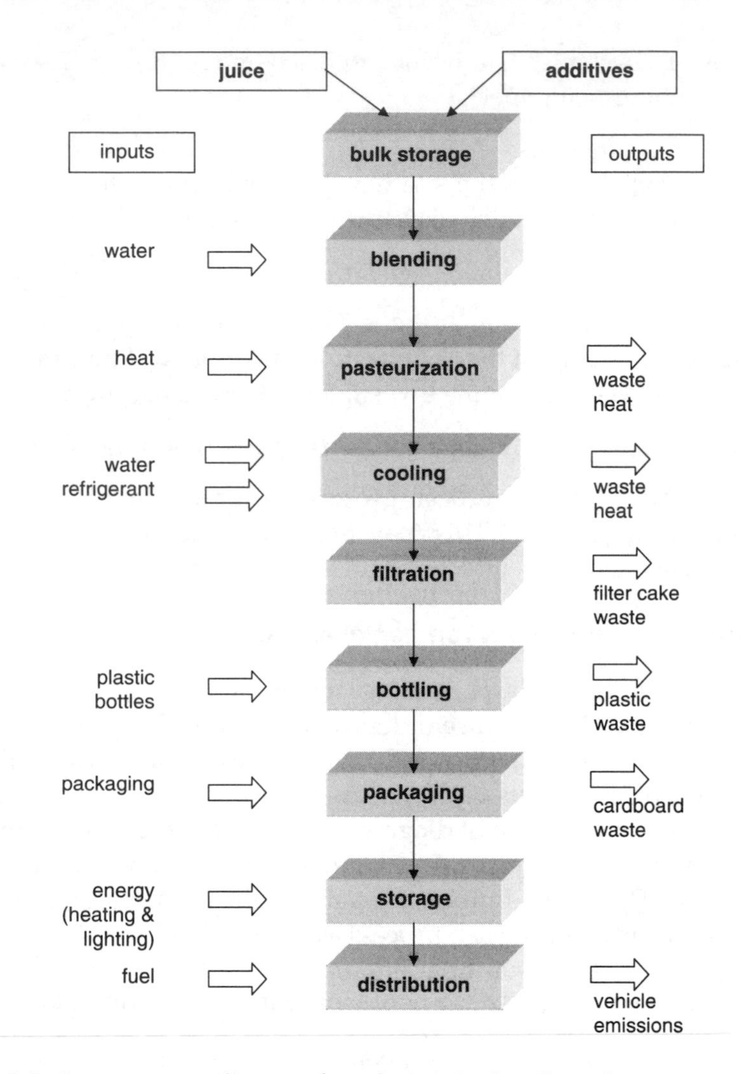

Figure 2.2 Input-output diagram for a fruit juice bottling plant

- Abnormal conditions generally include unexpected events and start-up and shut-down of continuous processes.
- Emergency conditions include fires, floods, chemical spillage and fugitive emissions of gases.

Therefore, typical questions to ask are:

1 Has a formalized risk assessment been carried out?

One method of assessing risks is to use the concept of:

Source ⟶ Pathway ⟶ Target

And that for a risk to be significant, there should be no breaks in the above links or chain. An illustration of this could be the example of a storage area for waste oil drums (the source) being positioned at the top of a grassy slope leading down (the pathway) to a small river (the target).

Clearly, in the event of a spillage all three components are linked and the risk of pollution exists.

Consider the same example whereby the waste oil drums (the source) are now double bunded removing the central link of the pathway, and thus the risk of pollution is less.

2 Are there emergency plans in place – especially concerning a major spillage or fire?

3 Are staff trained in operating to such plans?

The outcome of the inputs and outputs exercise and the review checklist can be in the form of a report to the senior management of the organization.

The preparatory environmental review report

Figure 2.2a is an example of a summary report produced for senior management and highlights priorities for action, areas which need improving to prevent potential issues such as complaints and other issues which represent good or best practice.

Clause 4.1: General requirements

This clause of the Standard states that the organization shall establish, document, implement, maintain and continually improve an EMS, the requirements of which are described in the whole of clause 4. (Clause 4 actually encompasses all the requirements of the Standard!)

Preparatory Environmental Review Report Executive Summary

Conducted on 24/1/2004

Key: P = priority
C = possibility of environmental complaints
G = best or good practice

	Emissions:
P	Comply with the requirements of the coating process permit.
P	Maintain documented monitoring requirements of visual and olfactory survey as required by the permit to operate.
P	Compile register of legislative and regulatory requirements.
C	Continue to research and review potential use of alternative coating processes to reduce VOC emissions by using water-based paints and inks.
C	Develop codes of practice for sub-contractors arriving on site for the proposed extension to the factory to reduce fugitive dust emissions.
G	Monitor and control external noise emissions and intrusive lighting as appropriate, to reduce disturbance to adjacent sensitive amenity areas.
	Discharges:
P	Ensure continual compliance with trade effluent consents to discharge.
C	Determine the precise route and discharge outfall of the surface water drainage system in order to establish if there is a pathway for solvent spillages from the main chemical stores.
C	Ensure that all site drains are mapped and covers to the surface water and foul sewer are painted different colours to aid identification.
G	Carry out regular inspection and maintenance of surface water interceptors and draw up an operational control to manage this.
	Waste management:
P	Monitor and ensure regulatory compliance with regard to maximum allowable storage requirements of dry wastes.
P	Ensure all waste disposal facilities are appropriately licensed.
G	Clearly label all waste collection areas to assist in the segregation of waste.
	Storage facilities:
P	Establish whether bunds at the tank farm have 110% capacity.
G	Provide adequate labelling of hazardous raw material stores.
G	Investigate feasibility of storing chemical drums under cover to prevent water ingress and reduction of risk of damage by vandals.
G	Perform regular inspections and maintenance of bulk storage tanks.
G	Perform an energy audit and draw up measures for heat conservation in warehouses.
G	Limit the variety of packaging used to facilitate easier segregation of waste.

Figure 2.2a Preparatory environmental review report

P	**Suppliers:** Increase the environmental awareness of key suppliers through questionnaires and assistance with ISO 14001 implementation where appropriate.
G	**Customers:** Provide customers with environmental performance criteria of finished products; increase their understanding of their duty to inform end users of the safe and appropriate disposal at end of life of products.

Figure 2.2a (Continued)

As discussed in the previous chapter, this clause of the Standard is deceptively brief. However, referring to the related part of the Annex within the Standard (A.1 General Requirements) several fairly wordy, descriptive paragraphs are used to explain that the intention behind implementation of the Standard is to give improvements in environmental performance. By the continuous process of reviewing and evaluating, an EMS will be improved with the intended result of improving its environmental performance.

There is no real guidance as to how mature an organization's EMS needs to be to qualify for certification. The phrase 'establish and maintain' does not make this clear. For example, how long must the system have been established, and how much documentation needs to be generated to demonstrate a history of implementation, before the third party assessment? Further guidance is given within Accreditation Criteria (see Appendix II).

Clause 4.2: Environmental policy

The Standard requires that top management shall define the organization's environmental policy and ensure that it:

- is appropriate to the nature, scale and environmental impacts of its activities, products or services within the defined scope of the EMS;
- includes a commitment to continual improvement and prevention of pollution;

- includes a commitment to comply with applicable environmental legislation and regulations, and with other requirements to which the organization subscribes;
- provides the framework for setting and reviewing environmental objectives and targets;
- is documented, implemented and maintained and communicated to all persons within the organization;
- is available to the public.

The policy should, therefore, be relevant to the significant impacts of the organization and should focus on them. For instance, a plastics processing company should not focus its policy on saving raw materials by recycling polystyrene drinking cups in the canteen yet ignore highly significant waste from the manufacturing processes.

There are conflicts that need to be considered when producing a policy statement for any organization and a balance must be achieved. On the one hand, the policy must be specific, yet general enough for the public (remember that the Standard requires public availability) so that a 'lay' person could read it and identify the processes/products of the organization and what the organization is planning to achieve with respect to its environmental issues. On the other hand, the policy must not be so specific that it becomes outdated too quickly, or that the objectives and targets are too exacting and hold the organization to promises it cannot keep.

For some 'blue chip' organizations whose 'brand name' is so strong and uppermost in the public's eyes, they argue that there is no need to be too specific in detailing their activities/products/services. The consumer or stakeholder is well aware of what they offer. Further, if the organization is somewhat larger, perhaps a division of a multinational, the policy may be drafted by the corporate legal and/or marketing departments on an international basis. Such policies will take some time to come to fruition, requiring many inputs until a balanced statement is achieved. As such these policies may be couched in bland terms to cause neither offence nor commitment at government and investor levels and will certainly not be ammended or changed lightly. There is no harm in this approach to formulating policy statements but it is well to be aware of the

Standard's requirements in terms of 'appropriate to nature and scale of the business'.

One solution adopted by some organizations is to produce a meaningful policy but without details of objectives and targets. A more detailed document is available separately upon request, via a different document which is perhaps reviewed every 3 to 6 months whilst the policy withstands scrutiny from year to year. The organization should have nothing to hide and everything to gain by demonstrating environmental responsibility.

Commercial confidence must also be considered during policy writing and it was never the intention of the Standard to compromise an organization by forcing it to reveal sensitive information that could be used by competitors.

As a guideline, the certification body is looking for a balanced statement that can be audited against. Therefore, if an organization states that every member of staff and all contractors have been trained in environmental awareness, the auditor will look for evidence of this.

A good guideline is to strike a balance in the policy. Include longer-term and short term objectives as well as highly specific and broader objectives.

Where an organization is part of another organization or corporate body, it should ensure that its site environmental policy does not conflict with any statement in higher-level policies of the wider organization – and yet be very relevant to the site. The Standard focuses on this requirement to ensure that no higher-level policies would prevent the lower-level policy being carried out effectively at the site level.

If, for example, targets are set for a reduction of a particular emission to atmosphere, the auditor will need to see evidence of such reductions. The decision-making processes used by management in setting particular targets will also be examined.

Figure 2.3 illustrates a rather weak environmental policy that raises more questions (as in the notes at the side) than it answers. Apart from not including elements required by ISO 14001, could it actually be understood by the stakeholders? Could it be audited against by both internal and external auditors?

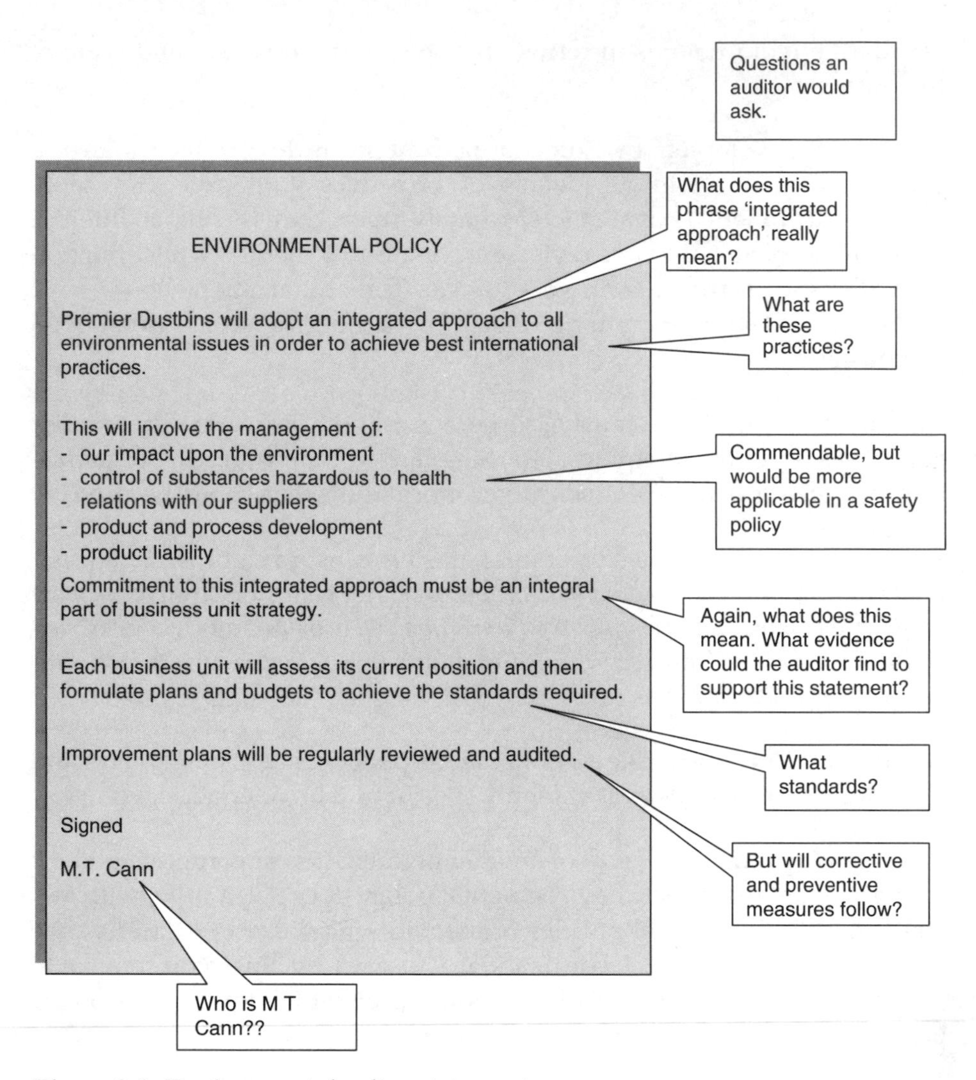
ENVIRONMENTAL POLICY

Premier Dustbins will adopt an integrated approach to all environmental issues in order to achieve best international practices.

This will involve the management of:
- our impact upon the environment
- control of substances hazardous to health
- relations with our suppliers
- product and process development
- product liability

Commitment to this integrated approach must be an integral part of business unit strategy.

Each business unit will assess its current position and then formulate plans and budgets to achieve the standards required.

Improvement plans will be regularly reviewed and audited.

Signed

M.T. Cann

Figure 2.3 Environmental policy statement

This policy does not address all the requirements of the clause of the Standard! It does not mention compliance with legislation, commitment to pollution minimization and does not appear to relate to a preparatory environmental review. The organization manufactures dustbins yet the policy does not reflect the particular environmental aspects and impacts of dustbins. Finally, who is M. T. Cann?

A more balanced policy statement by *Premier Dustbins* is illustrated in Figure 2.4.

If, for example, targets are set for a reduction of a particular emission to atmosphere, the auditor will need to see evidence of such reductions. The decision-making processes used by management in setting particular targets will also be examined.

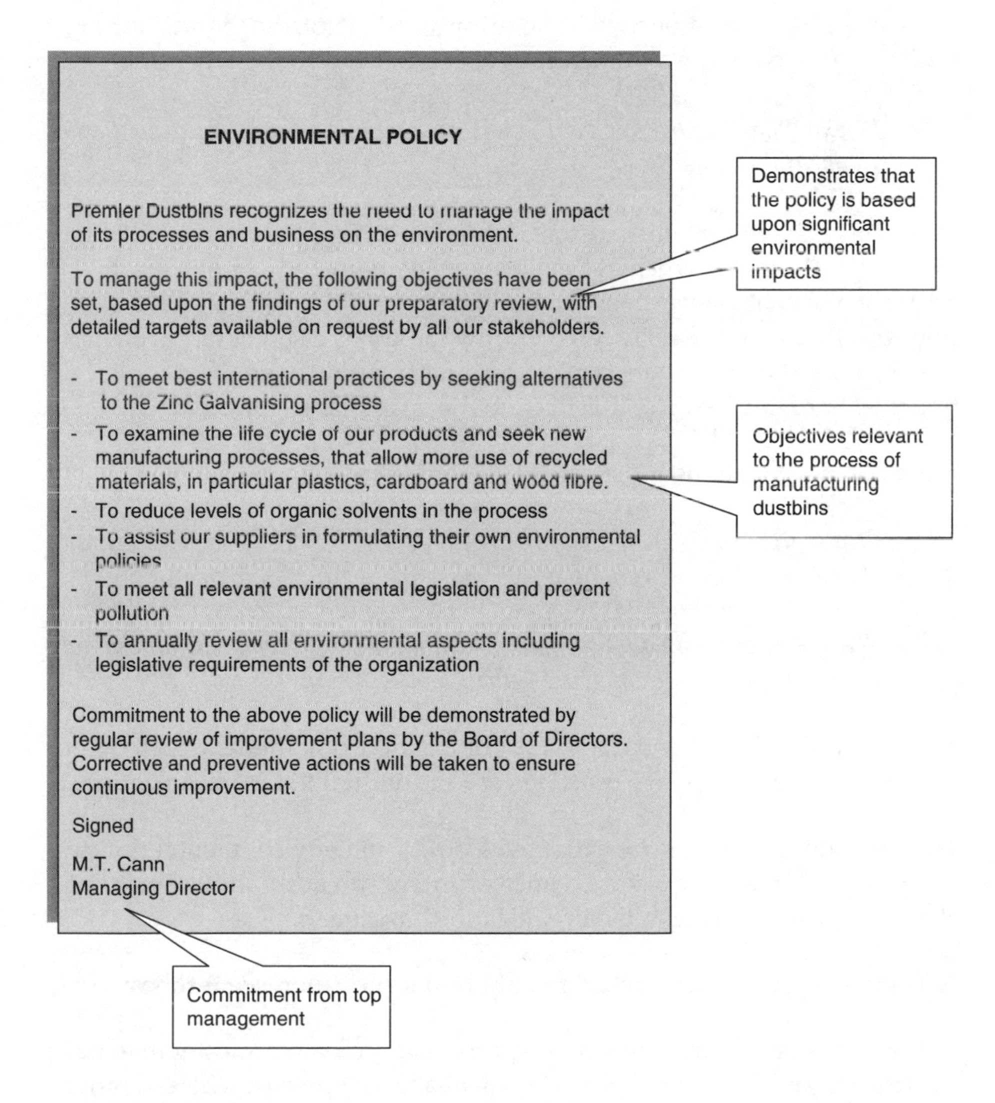

ENVIRONMENTAL POLICY

Premier Dustbins recognizes the need to manage the impact of its processes and business on the environment.

To manage this impact, the following objectives have been set, based upon the findings of our preparatory review, with detailed targets available on request by all our stakeholders.

- To meet best international practices by seeking alternatives to the Zinc Galvanising process
- To examine the life cycle of our products and seek new manufacturing processes, that allow more use of recycled materials, in particular plastics, cardboard and wood fibre.
- To reduce levels of organic solvents in the process
- To assist our suppliers in formulating their own environmental policies
- To meet all relevant environmental legislation and prevent pollution
- To annually review all environmental aspects including legislative requirements of the organization

Commitment to the above policy will be demonstrated by regular review of improvement plans by the Board of Directors. Corrective and preventive actions will be taken to ensure continuous improvement.

Signed

M.T. Cann
Managing Director

Figure 2.4 Revised environmental policy statement

Structure of the environmental policy

The following statements are extracted from other environmental policies and are examples of organizations not really considering the message they are conveying to the stakeholder:

> *'We will monitor and minimize the impacts of current operations on the environment.'*

Does this mean all impacts will be minimized – not the significant ones initially? This could be a tremendous task and may not be workable.

> *'We will ensure that sub-contractors and suppliers apply equivalent environmental standards.'*

Does this mean that all sub-contractors and suppliers have to obtain ISO 14001? If so, by when? How can this be enforced? How can this be controlled? What power does the implementing organization have over a sole supplier who can easily sell their raw materials elsewhere, particularly when losing the organization as a customer would cause the supplier no financial hardship? Again, this may not be a workable objective.

More balanced statements are:

> *'Through measurement and examination of the impact of its own activities the company seeks to eliminate or reduce the production of pollution.'*

> *'We will work with suppliers and Customers to influence them positively in terms of our environmental policy.'*

The policy should not end up as a list of isolated statements which do not have substance. The policy must be very individual to the organization.

The following examples are extracted from actual environmental policies. Such examples show a good balance of intent, breadth and scope whilst also being very specific to the organization's business.

A *lighting company* manufacturing light bulbs and fluorescent tubes:

- To consider the environmental aspects of all processes and/or materials, existing and new, and to apply best available technology, where appropriate, to help demonstrate our commitment to the prevention of pollution.

- To fund research into longer life products, thus saving valuable and non-renewable raw materials.
- The factory's objectives to reduce the environmental aspects on site will be made available to the public and stakeholders, through the improvement plan and associated projects. These will be implemented through individual improvement meetings.
- To consider the environmental policies of our suppliers when purchasing materials and to encourage the use of recycled and/or recyclable products where practicable.
- To seek out alternatives to the use of rare metals in the manufacturing process.
- To comply with all relevant environmental laws, regulations and corporate policies. The plant will work to achieve whichever sets the highest standard.

A *plastics processor* manufacturing packaging film:

- Within society, we will help to educate all consumers about the benefits and sound uses of plastics films and will be active in external initiatives to develop the efficient use of post-consumer waste.
- We will ensure that all internal scrap is reprocessed and made into saleable products.
- We will reduce virgin material consumption by using recycled or reprocessed material but without compromising customer and consumer quality requirements.

A *brewery*:

- We will minimize use of raw materials which have been grown using chemical agents (fertilizers and insecticides).
- We will only advertise responsibly. Campaigns will not be directed at vulnerable members of society.
- We will examine and reduce distribution costs by increasingly using haulage companies who are considering environmental controls, such as software route optimization and use of 'green' fuels.

A *frozen food manufacturer*:

We will focus on the following environmental issues:

- Reduction of energy costs – by funding research into improved insulation technology.
- Reduction of distribution costs – by consideration of transport routes, types of vehicles, fuel type, use of speed governors and driver training.
- Use of local suppliers of produce, where possible, to reduce transport costs.
- Reduction of food wastage by better shelf-life management.
- Use of brownfield sites rather than greenfield sites for all new freezer depots.
- Formulation of a purchasing policy on seafood derived only from sustainable fish farming practices.

A *city public authority*:

- Where possible, we will encourage new businesses in the area to use brownfield rather than greenfield sites.
- We will phase out the use of organic weedkillers used in the borough's green amenity areas such as parks and football fields.
- We will rationalize our fleet of commercial vehicles, using the most efficient vehicle for the job and, where possible, use electric vehicles.

A *forestry organization*:

Our environmental policy includes a commitment to a programme of sustainable forestry management:

- To preserve the natural bio-diversity of life within the forests.
- To use planning in all felling and transportation systems to minimize disturbance to the environment.
- To increase hardwoods by modifying forestry practices.

Finally, any organization should ask itself what it is attempting to achieve by the wording of its environmental policy. Minimum compliance with the Standard only? Or can the wording be such as to go one step further?

Thought should be given as to what evidence can be presented to the certification body to demonstrate 'commitment'. The environmental policy must demonstrate 'commitment to the prevention of pollution' (Clause 4.2(b)) and 'commitment to comply with relevant environmental legislation' (Clause 4.2(c)). Although objectives and targets will be audited by means of achievement against measurable targets, the auditor, for example, needs to gauge commitment and in many ways this can only be verified by interviews with top and senior management rather than by reviewing evidence contained in files or in records. The interviews will test whether the EMS is seen as an integral part of the business management functions and whether senior management actually mirrors their actions. The auditor will rely upon his experience and intuition to gauge this somewhat intangible label of commitment, and so the company should ensure that top management can indeed present their case.

It can also be said that although the interview technique is a strong tool to test commitment, tangible evidence will also be sought. The auditor will be mindful of such things as:

Housekeeping: itself is not a measure of protection of the environment but certainly a good indicator of management commitment and communication across the whole site.

Communications: if noticeboards, team briefs, site intranet are devoid of all things environmental it sends a message that the EMS is not all that important and perhaps only an elite group know all the details.

Environmental targets: could well be set at the minimum necessary to meet legislation. Targets set above tend to show 'commitment' to genuine improvements.

The policy also needs to be publicly available. However, this should be viewed as a positive requirement of the Standard and the opportunity should be grasped to provide more information to interested parties. The policy demonstrates that the organization does take its environmental responsibilities seriously and is taking specific constructive action to manage its environmental impacts. This can only improve the image of the organization in the eyes of the public and other interested parties.

Clause 4.3: Planning

This main clause addresses all the steps necessary in designing from the initial identification of environmental aspects to the setting of objectives and targets within the framework of applicable legislation.

Clause 4.3.1: Environmental aspects

The Standard requires that the organization shall establish and maintain procedures to identify the environmental aspects of its activities, products or services that it can control and those over which it can be expected to have an influence, in order to determine those which have or can have significant impacts on the environment. This planning also includes new developments, activities, products or services. Influence refers really to suppliers and customers i.e. the indirect aspects expanded upon later.

For many organizations, the key difficulty at this stage is the subjectivity associated with the identification and evaluation of environmental aspects. Since this exercise is the fundamental element in establishing a relevant EMS, then the methodology used, and the judgement exercised in assessing the significance of impacts is of crucial importance. Of equal importance is the discernment shown in being aware that an environmental aspect can be considered to be insignificant.

The screening mechanism – that is, the criteria for significance used to separate out the significant from the not so significant – must be reasonably transparent. This is because it will probably contain an element of subjectivity – corporate and personal opinions, perspectives and prejudices.

One thing is certain. There must be a range, or differing levels, of significance because clearly, if all environmental impacts are of the same significance, then none are significant! There has to be one impact (or perhaps two or three at the most) that has the highest 'significance', or shares the highest significance. However, by way of a slight 'blurring' see 'significance rating models' in the next section).

This is without doubt the most important part of the Standard. All the other elements are linked to this fundamental concept. It is the area where the implementing organization must spend the most time. Indeed, it is also the area where the certification body spend a significant amount of time examining and auditing.

Thus for each individual organization, the possibly long lis
aspects identified during the preparatory environmental revi
to a more focused list of the most highly significant environmen

How to identify environmental aspects

Much of what follows will have been performed during the PER. How er, the PER is generally a 'one off' audit and environmental aspects identification must be a dynamic process. The approach can be used of course for any new product, process or service that the organization is considering but the organization must first understand what aspects of its activities may cause adverse change. The main causes of change to the environment fall into the following groups:

- Use of resources such as energy, raw materials
- Releases into the atmosphere from normal activities – dust, noise, heat, odour, waste
- Accidental releases to the environment – such as fires (smoke and toxic gases) and leakage of chemicals, solvents or fuel (diesel)
- Development of land, including visual impacts, landscaping, drainage, pest control, changes to natural habitats
- Products and by-products

Having identified what activities may cause changes to the environment, it is then possible to determine what those changes will be.

As an example, a plastics company, using a wide variety of plastics such as PVC, PET, polyesters, polyurethanes and nylons should gather inputs from all its internal experts (personnel who have a wealth of experience of their area of activity within the organization). A good area to start would be with raw materials – plastic powders and granules – because their consumption will be the highest in the manufacturing process. Each plastic type will have associated environmental aspects which can be identified by asking the following questions:

1 Is it derived from a non-renewable resource?

2 Does it require a high input of energy (heat) for it to melt and be moulded into a finished article?

3 At the end of its life, is it capable of being easily and economically recycled?

4 Does it have undesirable by-products during moulding (gases, particulates etc.)?

Additionally, there could be a wide range of ancillary raw materials.

Each one will have its own individual environmental aspects which must be evaluated for significance. Clearly, this process requires a methodological approach with the collection of data for the consumption of materials, production of scrap, waste sent to landfill and so on, to enable an objective evaluation for significance.

The environmental management representative should act as co-ordinator here, to ensure that each individual, each section, each department, each process gathers this information. Certainly there will be individuals who will have an immediate contribution, because they have been employed in the organization for a number of years and probably have been collecting such data. Some of this information may be subjective but nevertheless will be useful.

For example, over a period of years the business will have been subjected to the requirements of regulations and legislation. There will be personnel within the organization who will possess knowledge that is relevant to particular legal obligations. Such knowledge may of course be superficial. An employee may have been responsible, over a period of time, for completing the necessary legal documentation related to solid waste being sent to a landfill site. Nevertheless, this experience is valuable and should be utilized by the organization. Clearly, the message is to use the existing knowledge within the organization.

How to evaluate environmental aspects

The Standard places much emphasis on the word 'significant' and the judgement of 'significance' is a critical issue, which bears upon a fundamental conflict between, on the one hand, the need to ensure that important aspects are not overlooked by cursory assessment and, on the other hand, the need to pay attention and assign resources (in a responsible manner) to those aspects which are truly important. The difficulty is exacerbated by the absence of any universal measure for comparative assessment of widely different environmental impacts.

ISO 14001 does not give any detailed guidance on this recognized difficulty. It is a generic standard. Because of this lack of prescriptive guidance, some application guides were developed by trade associations to assist members in this area.

For example, some sector guidance may advise on certain technologies or practices to be avoided – or certain raw materials. One guide, written for the chemical industry, advises that emergency response procedures should include media control. During major chemical spillages this takes on a major significance. Textile organizations are advised to change from bleaching processes using chlorine to less environmentally harmful, peroxide processes.

No definitive style or approach is in evidence throughout the guidance, having being developed by different writers and committees and this may be an area requiring control and standardization in the future. There are no rights and wrongs for aspects evaluation and a number of models for rating of significance are described as follows.

Models for the rating of significance
A) Risk Assessment Model 1

Risk assessment, which has been a management tool for many years (especially in the health and safety field), has two components:

1 The likelihood (or probability) of an incident

2 The likely consequences of that incident (or gravity of the incident)

Using a simple analogy, consideration of the 'risks' involved when travelling by commercial airliner will tell us that the likelihood of an incident (accident) occurring is very low, but the consequences of that incident are severe (often fatal).

By comparison, the risks of having an accident when travelling by car show us that the likelihood of an accident is higher, but the consequences are usually minor or trivial (broken limbs, cuts and bruising, scratches, shock etc.).

So, if the probability of an incident is low (because of procedures within a management system – be it health and safety, or environment – and/or technology) but the consequences are high (pollution, prosecution, loss of public image and market share) then the risk is still an item to be considered as having some significance.

There is some fairly sophisticated software available to assist in aspects evaluation; these computer programs have their uses but they invariably rely on some subjective decision-making by the user. A basic approach that can assist the process of evaluating risk assessment is to construct a simple matrix:

Likelihood of occurrence		**Consequences of that occurrence**		**Significance**
low	×	low	=	insignificant
high	×	low	=	moderately significant
low	×	high	=	significant
high	×	high	=	highly significant

Thus only the product of 'high × high' will score as highly significant. This matrix has its merits initially but the organization is encouraged to develop a more sophisticated model such as described below.

A table can be drawn up with the environmental aspects of the organization in the vertical column, and the criteria against which the aspects are assessed along the horizontal axis (see Figure 2.5). It is at this stage that a simple scoring system needs to be used. The advantage of using a scoring system is that numbers, and the products of those numbers (or even simple addition of those numbers), are easy to follow. It is easier to form a mental picture of what is being achieved using numbers than it would be using words. It is only a mechanism for allowing us to simplify a complex undertaking. So instead of having an order of significance starting at 'extremely significant' and going to 'highly significant' then 'moderately significant' and ending up with 'insignificant', a system of numbers

Reference number:			1	2	3	4	5	6	7	8	9		
	Risk												
Aspect	Probability	Consequence	Product (i.e. risk)	Past incidents	Nuisance	Abnormal	Local/Regional/ Global	Time-scale	Future activities	Legislation	Lack of information	**Score**	Comments/actions
Emissions to air	3	3	9	2	2	2	2	2	1	3	0	23	New plant on order
Emissions to water	2	3	6	2	1	2	1	1	1	3	0	18	Site discharge consent
Solid waste	1	1	1	1	3	3	1	2	1	2	3	17	Require weight of wastes
Use of energy	1	1	1	1	1	1	1	1	1	0	0	7	Energy audit planned
Noise	1	2	2	2	1	1	1	1	1	0	0	9	
Visual impact	1	1	1	1	3	2	1	1	1	1	1	12	Several complaints
Ecosystems	1	2	2	1	1	1	2	1	2	0	0	10	Small stream at boundary of site
Transport	1	1	1	1	1	1	1	1	1	1	0	8	
Suppliers	1	1	1	1	1	0	1	1	1	1	3	10	Questionnaires sent out

Figure 2.5 Register of environmental aspects

between 1 and 10 (with 10 being the highest significance) is far easier to manage and understand.

Figure 2.5 uses a key of:

***1** = low significance*

***2** = medium significance*

3 = high significance

The following nine criteria are cross-referenced in the figure:

1 Risk to the environment. In terms of the nature of the hazard, the probability of its occurrence and the likely consequences, should an incident occur.

2 Past incidents. The numerical scoring system allows more scope for looking at the history of operation of a process. For example, if the process is thought to be well controlled but there have been problems of an environmental nature in the past, then it could be argued that the likelihood of an incident is higher compared with the incident-free record of other processes. The site of the organization may have been contaminated by previous use and may always have the potential for a pollution incident.

3 Actual or potential nuisance. Is there any nuisance to neighbours – monitored, perhaps, by complaints? Is there any potential nuisance from a proposed expansion to the site?

4 Significance (and process conditions). Significance should also be considered in the context of normal, abnormal and emergency working conditions. Some processes have a much greater potential for pollution or even an environmental incident during start-up than when the processes are stabilized.

Note: Definition of abnormal

Several viewpoints are offered as below:

Definition 1:

One organization identified abnormal conditions as the requirement to work overtime. Overtime cannot be guaranteed to occur at set times. It will happen in response to urgent order requirements. This in itself may place no extra significant impacts upon the environment, but, if this overtime working involved weekend working, the authorized process monitoring or effluent plant operation may have to be modified by bringing in extra indirect labour (technicians etc.) to comply with legislation and ensure the potential for environmental incidents is not made greater by this 'abnormal' event.

Extra lorries may be required to transport raw materials and product. This would mean extra noise, lights on at night, etc. This may have to occur very early or very late at night. What impact would this have on local neighbours? This would need to be evaluated and dealt with.

Temporary staff recruited to deal with the extra work may not be entirely aware of some of the environmental responsibilities and so extra supervision would be required.

Definition 2:
Start-up and shut-down of processes can give rise to additional scrap being produced or some fugitive emissions, but it can be argued that this is not abnormal. For example, a manufacturing organization that starts up its plant on Monday morning and then shuts it down on Friday night is not performing an abnormal activity. It is an expected activity and occurs at regular intervals. The same could be said for annual shut-downs for maintenance. This again is an expected event and occurs regularly (albeit once a year). Both these examples will be planned activities with resources allocated. There may well be differences in the significance of environmental impact:

- Electrical machinery: may absorb more energy on start-up (to overcome the inertia of the machinery).
- Continuous process: the product 'quality' may fluctuate until a steady state is reached. There may be additional noise, dust, VOCS, etc. for a period of time.

However, the above can be reasonably predicted and the organization needs to evaluate whether such occasions would lift the evaluation of an aspect from being of low significance to a higher significance.

Definition 3:
Some organizations define the unexpected as the abnormal. Some unexpected events are of course emergencies and will be addressed later. (In ISO 14004, no real guidance is offered but there does appear to be a difference to be recognized by organizations between abnormal operating conditions and emergency situations.) However, abnormal events could cause undesirable environmental impacts or could increase the environmental consequences of an incident.

- Breakdowns: should be unexpected (given that an adequate amount of maintenance is performed). They could give rise to sudden oil leaks, dust emissions.

- Shortage of personnel: due to illness. (Vacations are not abnormal as they are planned). This could create a temporary shortage of skills in a key environmentally sensitive department or location.
- Vandalism: although the organization may have adequate security, casual vandalism may still occur with unexpected results.
- Adverse weather: untypical weather such as storms, floods, extreme temperatures may all contribute to raising a low significant impact to a higher status.
- Sudden customer demand: see above

5 Spatial scale. Another way of categorizing environmental impacts is according to their spatial scale. Impacts, both direct and indirect, may be divided into local, regional, national and global. Local effects are those occurring within the local vicinity of the cause. Regional effects are those which extend beyond the locality of the cause but are still spatially limited. National effects are clearly those which occur on a national scale and global effects are those occurring on an international scale.

6 Time-scale. An additional categorization method could be to consider the time-scale over which environmental impacts occur. Occasional impacts which have no acute, or long-term, consequences may be viewed as less serious than effects which are likely to occur frequently and which have serious and/or long-term consequences.

7 Future activities. What strategic plans are in hand for expansion of the business or for producing a new product line or production process?

8 Legislative requirement. If an organization's process is subject to legislation (an authorized process) or discharge consents, then it is argued that, by definition, these environmental aspects are significant – otherwise why would such legislation be in force? A third party is, in effect, telling the organization that the issues are significant.

The consequences of infringement could be severe: heavy fines, imprisonment for company officers, negative publicity, etc.

9 Information. If there is a lack of information on which to make a satisfactory appraisal, then this issue should automatically become a significant environmental impact, and stay at that status until further information proves it to be otherwise.

B) Risk Assessment Model 2

This uses a variation on the above and proposes a score based on:

1 Frequency of an event taking place

2 Quantities of material involved during that event

Thus it can easily be identified that a daily event of 20 tonnes of hydrochloric acid delivered on site, should be given a higher risk factor than a monthly delivery of 1 tonne of the same material.

Note

Care should be taken when using numeric systems, however, that the numbers derived make sense. An over-complex system may give a possible scoring range of 0–300, which is very unwieldy. If such a scoring system only generates numbers in the range 10–50 then there is no point in having such a wide range. A scoring range of 1–10 has been found to be adequate in practice: low significance receiving a score of 1 and high significance equalling 10, with the majority of 'moderately significant environmental impacts' clustered around 4–8. Thus, once the evaluation process has been performed, it makes sense to stand back and look at the numbers. For example, an organization may arrive at a range of numbers from its matrix, from 0 (insignificant) up to 100 (highly significant). However, the majority of scores may well cluster around 50 or 60 with differences of only 1 or 2 between. Given the amount of subjectivity involved, it may well be that choosing an impact with a score of 59 as being more significant than one with a score of 58 is an error and that the level of significance could in fact be reversed.

As with any scoring system, common sense must prevail and in areas which are 'grey' some subjectivity must be used. Provided that such subjectivity is equally applied throughout, then this will withstand scrutiny by the certification body.

Other organizations use a combination of letters and numbers to create an alphanumeric score. However, it is important to remember that the scoring system is not absolute – it is merely a means for the organization to make sense of a very complex set of environmental concepts and interactions. Even with these simple sets of numbers, some subjectivity is inherent and the final numbers should be examined and, perhaps, an element of 'weighting' put into the numbers themselves. This weighting may arise from circumstances peculiar to the individual organization. For example, proximity to an amenity area well-used by members of the public, may give rise to a higher number of environmental complaints than would be 'normal' for this particular industry sector.

C) The 'Learned' Committee Model

This is more likely to be adopted by a smaller organization who may have few significant impacts on the environment. It has its merits, but the important thing is to ensure that, say, in 5 years time, will the scoring system, or weighting system still give a consistent and valid answer to the question of significance? The original committee may have disbanded and new members may have different points of view and ideas on criteria for significance.

D) Internal Audit/Corrective Action Model

This model is based upon awaiting the outcome of internal audits and corrective actions to identify issues.

This could be seen as a 'reactive' process but providing the internal audit system is robust and that corrective and preventive actions are carried out effectively, then this has its merits in a smaller organization in directing limited resources to weak areas of the system.

E) All Aspects Significant Model

All aspects (negative ones) are listed and deemed to be significant. The rational is that they all have an adverse impact on the environment, and therefore they should be reduced. Therefore no scoring system is used. Instead, the organization sets time-scales for what they can do with their resources. Thus the management programme includes all aspects, but with a wide range of time-scales for action, dependent on resources available and the ability to influence, change and improve.

Care must be exercised when it comes to legal requirements however. It cannot be stated in the management programme that compliance with a certain piece of legislation will be deferred until the finance can be allocated to comply.

Summary

Each implementing organization will develop its own model. Provided that the reasoning behind the process is sound, the certification body will not find fault with it.

Points to consider during the evaluation process

Measurements
If environmental impacts cannot be evaluated by direct measurement methods (such as those used to measure waste such as volume and rate of discharge, content of heavy metals, acidity or alkalinity), it is recommended that calculations involving mass balances (inputs versus outputs of a process) or computer modelling are utilized.

Life cycle analysis
Implementing organizations debate about deciding how much of the life cycle of a product should they include when assigning significance to environmental aspects. There is no strict requirement in the Standard to perform a 'life cycle analysis' but if an organization has the resources to do this it can be beneficial. (See Appendix I for a definition of life cycle analysis.) A life cycle analysis may assist in future design changes and will demonstrate to the certification body an extra awareness of wider environmental issues. Standards are becoming available in this important area (ISO 14000 series).

Some organizations direct much effort into minimizing the future environmental significance of a product – by virtue of the design of the product – at the early stage of the life cycle. They argue that because of this, the effects on the environment during use by the purchaser, and the impacts on the environment at the end of the product's useful life, are minimal and not significant. Also, organizations want to know how far up and down the supplier and customer chain they should look for environmental probity. The life cycle analysis approach is to consider all the environmental impacts a product has, from 'cradle to the grave'. However, as stated, this is not specifically required by the Standard. Thus the focus here will be on practical steps to take with suppliers and customers.

Direct and indirect impacts
Environmental impacts can be categorized into two groups – direct impacts and indirect impacts. A direct impact is a change arising as a direct result of an activity under the control of the organization. An indirect impact is a change that arises as a result of someone else's activities; these activities are connected to the organization in some way but are less easily controlled as they can only be influenced indirectly.

a) *Direct impacts*
Direct impacts are the easiest to consider. Indeed, the majority of organizations will commence their environmental aspects identification in this area before considering indirect aspects. There is nothing wrong with this approach provided that evaluation of the aspects does not begin until all the indirect aspects are known. The reason behind this is that an indirect aspect may rank as being a highly significant impact, far outweighing any of the direct impacts of the organization.

Direct impacts are usually far easier to measure and monitor than indirect impacts and have been addressed in some detail in the preceding chapter.

b) *Indirect impacts*
Identification and assessment of significance of indirect impacts represents an altogether more difficult exercise. Although subjectivity was bound to be associated with direct impacts evaluation, there is even more scope for subjectivity when evaluating indirect impacts.

Although Chapter 1 only mentioned indirect aspects in terms of supplier and customer environmental behaviour the organization should also bear in mind some of the wider indirect environmental issues – issues of global sustainability, for example. Some of these issues lead to contradictions in terms of what an organization should do to demonstrate 'green' behaviour:

- Resource depletion versus lifestyle considerations – manufacturing packaging for gift-wrapping or greetings cards, adds little or nothing to objective, measurable improvements in the quality of life.
- Global warming – should an organization consider not using any fossil fuels and rely on manual labour for manufacturing?
- Product life cycle considerations – should all products be manufactured with built-in obsolescence? Good for continuing business reasons but wasteful of resources!
- Recycling and waste management – should all products be designed to be 100% recyclable or with minimal waste management requirements?

Clearly, the above points could be argued but are somewhat outside the scope of this book. Wrapping paper and greetings cards can enhance individuals' mood or attitude; fossil fuel consumption enables an organization to be competitive; 'built in obsolescence' tends to occur through inevitable advances in technology and designing products to be totally recyclable may incur unsustainable research and development costs and time-scales.

Some indirect aspects can almost be 'hidden away' in an organization's internal management policies. An example of a hidden issue is that of an organization that has a company policy of free fuel as a 'perk' to some of its managers. This is not really conducive to good environmental management if it transpires that such managers ignore the distance they travel to work every day and forego using public transport in preference to their cars. Here, indirect impacts are being created by the organization's policy. Encouraging use of public transport through ticket discounts, and so forth, could be a better policy option. In addition, moving from an old inner-city location to an out-of-town industrial or business park with purpose-built offices will probably save money on heating and lighting costs. However, generally such business parks have poorly developed public transport infrastructures and will force existing employees to use cars, thereby adding to traffic congestion and air pollution.

However, the intention of the Standard is to provide a focus on the most appropriate way forward. In practical terms this means only a consideration of those indirect environmental impacts closest to the organization – the impacts of suppliers and customers.

Suppliers and the supply chain

The start point is to identify all suppliers to the organization. This may well be readily available from the purchasing or accounting functions. This list of suppliers then needs to be reviewed to separate out those suppliers who have, in the judgement of say the environmental manager, environmental impacts, or potential environmental impacts, because of their activities, products or services that they supply. These suppliers are then contacted via phone, letter or questionnaire for further information.

This initial screening should include some common sense criteria. For example, a two-man window-cleaning service for cleaning windows of an office block on a weekly basis should not be worthy of a letter or a questionnaire whilst a weekly visit by a waste oil removal company should be.

The information received from these 'critical' or 'significant' suppliers can be analysed and reviewed for categorization of environmental risk. The following example allocates A, B or C as risk categories:

Category A Suppliers	Have significant impacts/or potential to adversely impact upon the environment. May or may not have an environmental management system
Category B Suppliers	Have lesser impacts – but need to improve their environmental controls
Category C Suppliers	Have no significant impacts and/or have an environmental management system

Having achieved a measure of environmental responsibility, probity and intentions of suppliers, the next step is to develop a purchasing policy which takes account of the above.

Develop a purchasing policy

An obvious policy, and one which is easy to devise, is to state that all category A suppliers (as above) will not be used in the future unless they obtain ISO 14001, or show measurable improvements in their environmental performance. This simplistic policy may not be so easy to operate. Some realities of purchasing must be taken into account. For example, a supplier may fall into the A category as above, but may have provided raw materials at a competitive price, to the required specification over a number of years and for commercial reasons will need to be used in the future. To have a purchasing policy that states no category A suppliers will be used, could damage the viability of the purchasing organization's market share.

Also if there is a sole supplier of a raw material, whose environmental policies are in question, (category A in the above system), then it is somewhat non-productive to state that this supplier will not be used in the future because the organization is going to have to purchase from them anyway.

Thus a practical and meaningful purchasing policy needs to be developed. Continuing with the above example this could be:

Category A	Seek alternative supplier as soon as possible if they do not react favourably to assistance to improve their environmental performance

Category B	Continue to use but monitor environmental performance via annual questionnaires of audits
Category C	Continue to use unreservedly

This is consistent with best practice which is to educate, train, assist and facilitate such suppliers to improve their environmental performance and reduce their risks to the environment. This could be done through a programme of offering training days for the suppliers management team, organizing seminars, offering best practice advice and so on. Additionally, plans could still go ahead for sourcing a more environmentally responsible supplier given all other purchasing criteria are equal. This in itself can be a longer-term objective within the EMS.

The following examples are illustrative of the above approach:

- A *chemicals manufacturer* is likely to exhibit high direct environmental impacts of both a polluting and a resource-usage nature. This organization will therefore have highly significant direct impacts which should be addressed first, before considering the indirect aspects of suppliers.
- A *chemicals stockist* will have some direct environmental impacts but of more significance will be the chemical suppliers (for example, indirect impacts). Stocked items need to be transported both to the stockist and the customer. Sub-contracted delivery services are an indirect environmental issue of significance. (The transport of an organization's products is sometimes overlooked. If the transport is sub-contracted then this is a purchased service from a supplier and this supplier is no different from any other 'materials' supplier.)
- A *plastics packaging manufacturer* will have some direct environmental impacts, such as workplace atmospheric pollution, workplace and neighbourhood noise, pollution and high electrical energy usage. However, there will also be some significant indirect impacts. These indirect impacts arise because the plastics supplier will be using non-renewable resources (plastics and energy derived from mineral oil) and the end-user, the customer, will be disposing of the plastics packaging. As packaging film is, by its very nature, destined for a short life followed by disposal, the indirect impact is highly significant.

Another step an implementing organization can take is to measure the level of awareness of environmental issues achieved by its suppliers. This information can be gleaned from very careful scrutiny of the answers obtained from questionnaires. For example, by careful phrasing of the questions in these questionnaires, an insight into a supplier's environmental awareness will be gained. The objective could then be to raise awareness by visits, offers of advice, or the inclusion of suppliers in your internal audit schedules. An organization should be able to measure awareness increase by sending a second questionnaire – perhaps 12 to 18 months later – and compare these replies with the previous answers. (See Figures 2.6 and 2.7 for example.)

Some organizations' purchasing policies may be dictated to them by the corporate body. This body may well be located overseas and operate to a different agenda or timetable of global environmental purchasing policy. The corporate body may well dictate that only named suppliers are to be used to gain the advantage of lower prices by purchasing in bulk.

This may conflict with the local purchasing policy of the ISO 14001 implementing site and is obviously a cause of concern. The way forward here would be to establish an objective to influence the corporate body, perhaps to gather evidence that the named suppliers are acting in a way that does not demonstrate environmental responsibility.

Customers – can they be influenced?

To influence the environmental behaviour of customers, is without doubt, a difficult requirement to meet. After all, an organization cannot dictate how a customer should behave – they will probably lose that customer! Some good practices are offered below. The amount of work necessary can be reduced by considering once again the concept of significance. An organization should establish from its customer database which customers are the largest in terms of volume of product they take, and which customers take products that have the most potential to impact on the environment. These customers will then be the first to be targeted by the organization, when managing its indirect environmental impacts.

The question of how far influence should be exercised along the customer chain can be answered in terms of significance. It may only be necessary initially to include just the immediate distributor of an organization's products. For an organization supplying replacement brake shoes for vehicles, it would be totally impractical if it attempted to influence every end user of its brake shoes (that is, garages and the general motoring public), to dispose of the worn-out brake shoes in an environmentally friendly way and

TO:
Company Name:

Address:

Representative for environmental issues:

Product or service supplied:

If you have been certified to ISO 14001, please enclose a copy of the certificate. The rest of the form need not be completed. Please sign and return:

Signed: Position: Date:

What do you believe to be the environmental impacts of your business activities?

Do you carry out environmental audits?

Do you have a written environmental policy?

Do you operate under any legal permits, consents or authorizations?

Have you ever been prosecuted in a court of law for breaches of environmental legislation?

Do you segregate/recycle any of your waste?

Do you have energy reduction programmes in operation?

Do you have any emergency plans in place?

Do you intend to obtain ISO 14001 certification?

Figure 2.6 Environmental performance questionnaire

recycle where possible. Indeed how could it check whether such influence was working? A more practical approach is to attempt to influence either corporate bodies such as local authorities (who will purchase replacement brake shoes in large numbers for their fleets of lorries, buses, snow

TO:
Company Name:

Address:

Representative for environmental issues:

Product or service supplied:

The person who fills this in should carry a reasonably high level of authority within the organization.

If you have been certified to ISO 14001, please enclose a copy of the certificate. The rest of the form need not be completed. Please sign and return:

Signed: Position: Date:

What do you believe to be the environmental impacts of your business activities?

The answer 'we do not pollute anything' does not answer this question!

Do you carry out environmental audits?

Do you have a written environmental policy?

If yes, who was it written by? It should be top management rather than say, office manager.

Do you operate under any legal permits, consents or authorizations?

The answer 'don't know' indicates a potential for breaches of environmental lawsl

Have you ever been prosecuted in a court of law for breaches of environmental legislation?

Do you segregate/recycle any of your waste?

Do you have energy reduction programmes in operation?

'Posters on office walls for switching off lights' may not be appropriate in an energy-intensive factory.

Do you have any emergency plans in place?

Do you intend to obtain ISO 14001 certification?

Figure 2.7 Second environmental performance questionnaire

ploughs, etc.) or the larger distributors who distribute to garages and spare parts stockists. By influencing a few distributors, a significant number of used brake shoes could be recycled or disposed of in an environmentally responsible way.

What form can the influence take?

- Continuing with the example above, the organization might fund research into alternative ways of producing brake shoes without use of asbestos.

- For an organization printing telephone directories (which have a limited life due to the constant number of changes) a financial incentive to return obsolete directories to a central collection point for recycling could be implemented.

- The manufacturer of plastic packaging could state on the packaging itself what type of plastic it should be recycled with. This may encourage some customers to demonstrate environmental responsibility.

In conclusion, it should never be the objective of an organization to make all suppliers and customers obtain ISO 14001. For some smaller suppliers in particular, the task of implementation may be very costly for them and achieve very little – their environmental impacts may not be appropriate or significant when compared with larger suppliers. Only the suppliers and customers who have a significant impact should be targeted for ultimate compliance with ISO 14001 – if this is thought to be the best way forward. Certainly, the history of quality assurance has shown us that smaller organizations have struggled to comply with the requirements of the Quality Standard ISO 9001 and have designed inappropriate systems. It was never intended that ISO 9001 compliance be 'forced' upon smaller organizations – who may have no practical need for such a system. The same is true for ISO 14001.

'Outside of scope' aspects

There are examples of divisions or sites belonging to larger multinational organizations, who have 'stand alone' ISO 14001 certification. Thus a small manufacturing site may produce a product in its basic form which is then shipped to other sites for packaging and distribution.

These packaging and distribution processes will not be carried out under ISO 14001 disciplines, but the small site as above will have little or no power to influence them directly.

This can be addressed by implementing longer-term strategies to influence the parent organization such as high visibility of the advantages of their ISO 14001 system at every available opportunity.

Clause 4.3.2: Legal and other environmental requirements

The Standard requires that organizations shall establish and maintain a procedure to identify and have access to applicable legal and other environmental requirements to which the organization subscribes, and to determine how they apply to its environmental aspects.

Examining the 'other environmental requirements' first, these might include a situation where, for example, the corporate headquarters of an organization decrees that certain solvents will be banned from all sites by a certain date. Clearly this directive must be obeyed and reflected within the overall environmental policy, programme, objectives and targets. As a further example, the use of the plastic PVC is under active environmental scrutiny because it is argued that, as PVC is a chlorinated material, it can have adverse health effects on humans and wildlife. Although such evidence is inconclusive, an environmentally responsible organization may mandate that all of its sites phase this material out until there is further research into the environmental impact and safety hazards of PVC.

Similar codes of practice will operate within other industry sectors. Local by-laws may also be applicable and these can be obtained from local authorities or utility companies. Most countries have government environmental agencies who will assist with national legislation information. The legal requirements will, of course, be considered by many to be the most important due to the possibility of fines, imprisonment or the forced closure of the business. Therefore, keeping up to date with legislation is important. However, bearing in mind the extent of the environmental legislation emanating from regulatory bodies throughout the world, this is no simple task.

For most organizations, it will be necessary to rely upon assistance from either their environmental consultants or the specialists who collate and publish details of legislation. These tend to be trade associations, law firms or specialist publishing companies. Some sources are listed in Appendix III Additional information. These sources will supply information of a general nature which will be relevant to the organization's activities.

It is highly likely that in some organizations, legislative knowledge is already possessed by certain individuals. For example, a transport department may well have records of waste sent to landfill and the necessary legally required paperwork and regulations to hand. The production manager may well know the requirements regarding discharging only certain consented amounts of process water to sewer. However, such knowledge may be fragmented and it therefore makes sense to collate information concerning applicable legislation in the form of a file or a register (see Figure 2.8).

***Notes on legal issues*:**

The challenge in a book of this size, which must retain its title of 'handbook' is to only include that which is meaningful and to act as a pointer for the reader to more detailed texts and more local environmental legislation. The other challenge is to ensure the information will not become quickly

Act / Directive	**Reviewed on (date):**	**Applicable to the site?**	**Comments**
Asbestos at work act 2003	23/10/02	yes	Site survey for presence of asbestos required
Oil storage regulations 2000	12/08/00	yes	Fuel tank of diesel – 3000 litres within 10 metres of canal at rear of site.
Landfill Directive 1998	01/12/98	no	no special wastes removed from the site
Wildlife and Countryside Act 1981	10/04/04	no	This act protects SSSI's from disturbance and development. The Company is in a centre of a city and no SSSI's.
Water Resources Act 1991	10/05/03	no	No water is used in the process. No liquid waste apart from domestic waste. No abstraction takes place.
Radioactive substances act 1993	10/04/04	yes	The Company has 2 radio active sources
94/62 EEC Directive on Packaging and Packaging waste	10/02/02	no	Currently amount of packaging is below the threshold value (50 tonnes per annum)
Town and Country Planning Act 1990	10/03/04	yes	The planned extension in 2006 will create noise. Some restrictions are placed on noise levels within the act
Environmental Protection Act Pt II Duty of Care	10/03/04	yes	Waste management procedures in place. Waste transfer notes kept and filed.

Figure 2.8 Register of legislation

out of date. Therefore the emphasis is on the principles of environmental legislation and guidelines for compliance.

Note 1) The Law
We may not always agree with laws – but we must obey them. Non-acceptance of a law, or pleading ignorance of the law is no defence when it comes to courtroom proceedings. For example, in the case of an adjoining river to a manufacturing site becoming accidentally polluted, it is no defence to state that the Company was unaware that this was an unlawful act. The legal system would maintain that the Company should have taken reasonable steps to find out beforehand.

Note 2) Penalties for breaking environmental laws
Certainly there are instances of heavy financial penalties (fines) for breaches of legislation. Additionally, senior management individuals have been sent to prison – usually in cases of proven negligence. In the UK, for example, legal guidelines exist with statements such as 'fines should be substantial enough to have a real economic effect on the organization', the purpose being to put pressure on managers and shareholders to comply.

Note 3) Concept of Strict Liability
Generally, offences under environmental law are subject to the concept of strict liability. As an illustration, consider the river in previous note 1, being polluted by a large drum of chemicals. If the reason for the incident is adverse weather conditions (storms blowing drums over causing spillage) or even criminals or vandals deliberately causing the spillage, then the Company is still guilty of pollution. All the Company can do is to plead mitigating circumstances which should reduce the severity of the fine. Such mitigating circumstances could be:

- Enclosed and locked storage areas for the drummed chemicals;
- Use of 24-hour security cameras and vandal proof fencing around the site.

This demonstrates to the court of law and the regulatory body, that due diligence was taken.

In a less dramatic scenario, if a river is polluted due to an interceptor not functioning – the organization is still guilty and will face a fine in the courts. If the organization concerned has a regime of weekly checks, inspections, audits under the umbrella of an EMS, then the level of fine may be reduced as the client was indeed taking reasonable steps and precautions to reduce pollution.

Note 4) Legislation – responsibility for compliance
It is the implementing organization's responsibility to ensure compliance with the law. This responsibility cannot be passed on to a third party, for example, the auditor from the certification body.

The external auditor, however, must exercise due diligence and by asking questions based upon his training, expertise, knowledge and experience, should discover if the organization is compliant.

Note 5) Determination of applicable legislation
Legislation is publicized by governments and regulatory authorities and should be readily available. Options can be access to internet sites, enquiries at government offices and reading of appropriate trade and professional journals.

In this way a register of legislation can be compiled. However, the register of legislation by itself is not sufficient. The client must demonstrate how each piece of legislation applies – or not – as the case may be with reasons for inclusion or non-inclusion in the register.

This register demonstrates that a review process has occurred by a responsible person. This process should of course be detailed in an appropriate procedure.

Legislation – the future
UK, European and international law is moving towards encouraging prevention of environmental problems rather than 'end of pipe technology' i.e. abatement mechanisms. A more holistic approach is becoming prevalent i.e. more sustainable preventive measures through BAT. In Europe, for example, the Landfill Directive places more incentives to produce less solid waste at source.

Clause 4.3.3: Objectives, targets and programme(s)

The Standard requires that the organization shall establish and maintain documented environmental objectives and targets at each relevant function(s) and level(s) within the organization.

Objectives should be the longer-term goals derived naturally from the environmental policy. It should be understood that each identified significant aspect will have an associated objective or objectives in some cases.

Quantification can then take place through measurement in order to meet such goals. Of course, all objectives and targets should be realistic (with rational decision-making behind them).

Objectives should be related to significant environmental impacts and can be couched in fairly broad terms i.e. to reduce energy use. Each objective should have a measurable target to demonstrate that the objective is being attained (or otherwise). Targets are more specific, more easily measurable detailed performance requirements which evolve from the objectives and allow an organization to verify whether the stated objectives will be achieved. Early warning mechanisms for targets not being met should be in place. The process of regular reviews and audits should address this adequately.

Quantification of targets is not always easy but sometimes there are suitable measures within the organization's existing management systems.

A broad objective could be to reduce waste by 10% compared to the previous year. One of the targets could be to reduce waste to landfill. Deciding how to measure this could be as simple as reviewing and analysing existing records within the organization (from waste transfer notes, weighbridge tickets and also from invoices for cost of removal, cost of landfill).

Setting reasonable and obtainable targets

A manufacturing organization:

Some targets will be dictated by the requirements of legislation and therefore are set from outside of the organization. That apart, using the example of landfill waste above, if an organization has identified from its preparatory environmental review that solid waste to landfill is a significant impact, what should be the target to aim for to reduce this amount of waste?

First and foremost, quantification of what is actually sent to landfill needs to be obtained. As stated above this could probably be obtained from weighbridge tickets, waste transfer notes and other records. If, from these records, it is discovered that in the previous year 100 tonnes were sent to landfill, how does the organization derive a meaningful figure for reduction? Is 1%, 10% or 50% reduction a reasonable figure?

Upon examination of these options, an improvement of 1% is meaningless as far as environmental significance is concerned. It will probably be difficult to measure this 'amount' with confidence. There is also the fact that the costs of the controls for this small reduction may outweigh any financial considerations – always an issue in any organization. The improvement of 50% would appear at first glance to be commendable and worthy of environmental attention, however, on closer examination it is probably somewhat unrealistic. Otherwise, why has the organization not done something about such a large improvement as this before?

Thus 10% appears to be a starting point and an achievable target – with measurable associated cost savings. If, as the system matures, this proves to be too difficult, then it could be adjusted to 8% or 9% as appropriate.

Obviously, only running a production line at 50% capacity, due to poor customer demand, will reduce waste by a roughly corresponding amount. Unless this is taken into consideration in the calculations, errors in the figures will occur.

A service organization:
Video conferencing has been used by many organizations to reduce the amount of travel by their staff between locations, and when used correctly will give real environmental, as well as commercial benefits. Ongoing fuel, hotel and meal costs to the organization are reduced, staff can be more productive (rather than driving), and of course the environment will suffer less from pollution from vehicle exhausts.

Cynics may well say that it has its uses, but will not stop senior executives flying off to the more exotic locations for conferences, seminars, sales meetings, etc. This may be true, but if it cuts down the time and expense of the mundane (80%) travel to less exotic locations, then there will be benefits.

The key is to measure by perhaps using a log to show which members of staff have used the video conferencing. Additionally noting the numbers of staff involved in the 'conference' and an estimation of the distances that could have been travelled – and now saved. Calculations can be performed to show fuel cost savings as well as costs of unproductive time spent travelling by staff that has been saved (of course, although less an environmental issue more of a business efficiency issue).

Management programme(s)
This clause of the standard requires that the organization establish and maintain a programme, or programmes, for achieving its objectives and targets. It shall include:

a) designation of responsibility for achieving objectives and targets at each relevant function and level of the organization

b) the means and time-frame by which they are to be achieved

As a further explanation of a), see Figures 2.9a and 2.9b and note that the example programme assigns responsibility to several individuals and not just the environmental manager.

'The means' and 'the time-frame' are also worthy of further explanation:

Means
The 'means' is the methods, methodology, the practical 'doing', employed to achieve the targets. Consider an organization that sets an objective to reduce fuel costs of company vehicles by say, 10% over a 12-month period. This is measurable (measured through the analysis of fuel bills), but unless the staff involved know exactly what they must do as individuals to achieve this, the objective may founder. Simply stating that 'staff will drive more carefully and avoid excessive speed and harsh braking' is quite meaningless in terms of being able to monitor and measure this. However, if a management programme is drawn up detailing the means then it could describe:

	'Means' or methods	**Actions required of staff**
1	The fitting of fuel economizers	Fitted to a select number of cars, and fuel consumption recorded by the drivers on a weekly basis.
2	Selected staff attending defensive driving lessons	Fuel consumption measured both before and after lessons to measure any difference. Drivers record in log book provided.
3	Purchase of software to plan journeys more efficiently	Drivers to note distance travelled and compare to the same journey travelled using the planning software.

Environmental Responsibility aspects	Significant impacts	Objectives	Targets	Means	Time-frame	
Cardboard, paper, filter cake	Filter cake contains heavy metals	Modify process to reduce heavy metals	Reduce by 50%		Q3 2004	Technical Manager
Energy – electricity, gas and oil	Potential for oil leakage from oil tank/costs of oil heating	Investigate costs of change to gas heating		Costing data in a report	Q3 2004	Facilities Manager
Nuisance noise from compressors noise from delivery lorries odour from process	Complaints from neighbours	Monitor site boundaries for noise levels		Perform on monthly basis	now	Environmental Manager
Redundant underground oil tanks	Potential for leak and ground contamination	Remove		Prepare plan of demolition	Q1 2007	Facilities Manager
Water consumption	Use of treated town's water	Test for borehole		Reduce towns water by 25%	Q4 2005	Environmental Manager
Discharge of effluent	Old pipework – potential for leakage	Check all pipework for integrity		Maintenance plan drawn up	Q2 2005	Maintenance Engineer

Figure 2.9a Management programme 2004–08 direct impacts

Environmental aspects	Significant impacts	Objectives	Targets	Means	Time-frame	Responsibility
Suppliers' environmental issues	Unknown	to influence	Unknown as yet	Issue of questionnaires	Q3 2004	Technical Manager
Customers' issues	Unknown	to influence	Unknown as yet	Set up meetings with customers	Q3 2004	Sales Manager
Sub-contractor issues – sub-contractors used for storage and distribution of product	Energy used in warehouse for heating and light	Reduce use of energy	Reduce costs by between 5 and 10%	Perform energy audit. Use a specialist organization	now	Environmental Manager

Figure 2.9b Management programme 2004–08 indirect impacts

Time-frame
Objective can have several associated targets, some may be 'tangibles' and others may be periods of time or dates. (Dates are very good as a means of measuring progress – in effect something either happens by a certain date, or it doesn't!)

In the above example, the economizer is either fitted to the vehicle by a set date, or it isn't. Only two situations are possible – both easily measured.

Thus a management programme, or programmes, can be drawn up and Figures 2.9a and 2.9b are examples. There is no set format for a management programme and Gannt Charts, spreadsheets or project planning software can be used.

Whatever the format, the programme should show the linkages, or relationships, between aspects, significant impacts, objectives, targets, means, time-frame and responsibility.

Clause 4.4: Implementation and operation

The Standard requires, via the following seven sub-clauses, the organization to put into place controls over all activities which have, or may have, a significant environmental impact. In effect, procedures will need to be implemented to ensure that the everyday environmental activities of the organization occur as planned.

Clause 4.4.1: Resources, roles, responsibility and authority

The Standard requires that, as a show of commitment, the organization shall provide resources – especially human resources – in order to facilitate effective environmental management. The successful implementation of an environmental management system calls for commitment from all employees in the organization.

The purpose of this sub-clause is to ensure that finance is available and that personnel are assigned responsibilities for part of the environmental management system and have a clear-cut reporting structure. Job descriptions, or project responsibilities from the management programme, may cover this requirement.

A management representative needs to be appointed. This can be (and is for the majority of companies) an existing member of staff who, regardless of other duties, has responsibilities for co-ordinating the activities of the environmental management system. There needs to be a direct authority linkage. For example, in the case of a potential environmental problem, the line of communication to senior management needs to be short so that action can occur readily. Commitment begins, of course, at the top level of management, but it is accepted that in larger organizations responsibility to be the management representative is often delegated to a less senior individual. However, in smaller organizations, the management representative might be the managing director himself.

Some organizations spend much of management time defining job responsibilities via documented job descriptions. Indeed, these job descriptions are an excellent method of addressing this clause of the Standard. However, care should be taken to describe the authority that an individual has in an emergency environmental situation. For example, an operator at a remote location may not have the authority to turn off a part of a production plant to prevent a spillage in an emergency situation (this would cause financial loss to the organization due to lost production). The operator may therefore need to refer to a higher authority for a decision which, of course, could lead to the incident becoming more serious due to the time delay.

It may well be that after the event, the operator could have made that decision and would have had the full support of management in dealing with the consequences of lost production (such as lost revenue plus, perhaps, the additional cost of nonconforming product requiring disposal). Thus such levels of environmental authorization should be defined in job descriptions so that all individuals are aware of what decisions they can make, especially in an emergency situation.

Clause 4.4.2: Competence, training and awareness

The Standard requires that the organization shall identify training needs and this clause requires that all personnel whose work may create a significant impact upon the environment have received appropriate training.

Thus the organization must satisfy the following four criteria:

1 **Ensure that training needs are identified**:

This can be performed via appraisals. In most companies this is an annual event – at the very least for salary review purposes. From this, training needs will be identified and a plan of either internal or external training planned. All individuals will need some level of training in the requirements of the environmental policy and a background to the requirements of ISO 14001. Some individuals will need specific training in emergency response. Others may need their roles to be changed and defined. An internal quality assurance auditor may well need to be 'converted' to an environmental systems auditor via an external training course.

2 **Ensure that these planned needs are met**:

There must be a system to ensure that such individual training plans are carried out as intended. Procedures will be needed to describe such mechanisms, as well as including a broader description of how the organization's training strategy is structured. In addition to specific external courses or seminars, internal workshops and briefings are an acceptable vehicle for training. Internal environmental newsletters are also part of the range of tools available.

3 **Verify that the training has achieved its purpose – increased awareness**:

This verification can be performed via feedback from training sessions: either a written report from the individual or a simple questionnaire to complete. Some organizations will ask personnel to undertake simple 'tests' to measure the effectiveness of the training.

Other ways of verifying 'awareness' could be through the internal audit system. Asking questions of personnel during such audits will give an indication of their knowledge.

One of the challenges can be the measurement of 'continuous improvement' within training. Clearly, measurable targets can be set for delivering the training, i.e. records of attendance would show who has attended such sessions. However, how can an organization 'measure' whether the knowledge (awareness) of personnel has 'improved' compared with their knowledge of say, 12 months ago?

The author can only suggest that the answers to written tests as mentioned above, be reviewed and compared to the answers of previous tests. Such comparisons should indicate heightened levels of knowledge, or otherwise, as a measure of improvement.

4 **Verify that following training, the individual is competent at applying the awareness gained to their particular job:**

This can be achieved by monitoring an individual's work, noting any improvements in work or, conversely, monitoring any persistent failure to absorb such training (for example, by not being aware of the consequences of departure from a specific work instruction).

Contractors, working on behalf of the organization, must also be subject to training requirements and this must be addressed in the training procedures. The employees of the contractor should have a certain level of training – that level to be determined by the organization. For both contractors and the organization itself, it must be kept in mind that the concept of significance must be applied to any training plans or programmes. The individual who is in an environmental front-line position – whose actions have the potential to cause a major impact on the environment – should have priority in environmental training followed by a more intensive scrutiny of awareness and competence than the individual whose actions have little potential to impact on the environment.

Further, if for example an individual is trained to operate a pH meter to check effluent pH, clearly both awareness and competence can be verified by giving the individual a sample of liquid (of known pH) and asking him to test it. If the results are correct it would tend to demonstrate that the training was successful.

Clause 4.4.3: Communication

This sub-clause refers to all types of communications, both internal, and external to, the organization. It requires organizations to establish and maintain procedures for internal communications between the various levels and functions of the organization. It also requires organizations to document and respond to relevant communications from external interested parties. Examples of internal communications include:

- Communicating environmental objectives and targets to employees
- Raising awareness of environmental issues to employees
- Communicating the environmental policy to employees
- Advising of nonconformances to relevant departmental heads

- Reporting incidents arising from abnormal or emergency operation to senior management

Examples of external communications include:

- Dealing with environmental complaints or proactively inviting such stakeholders, or 'green' pressure groups to the site. This could also involve inviting schools and colleges to the site for educational purposes.

- Responding to media enquiries, especially in the event of an incident.

An inability to communicate effectively within the first few hours of an incident could seriously reduce the company's ability to control the situation. This will undermine the company's reputation in the minds of staff, customers, the media and the public. (See also sub-clause 4.4.7 *Emergency preparedness and response*, later in this chapter.)

Clause 4.4.4: Documentation

The organization shall establish and maintain information, in paper or electronic form, to:

a) describe the main elements of the environmental management system and their interaction

b) include documentation required by ISO 14001

c) documentation determined by the organization itself

d) records required by ISO 14001

e) provide direction (or reference) to related documentation.

It is not the intention of this book to describe in detail how to write manuals and procedures to meet the requirements of the implementing organization and the Standard. In fact, there is no laid down format and each organization should develop its own style that it can work with. However, there are, within industry and commerce, certain styles (based upon years of experience) which are better than others. Meeting the needs of the organization and complying with the Standard is the first consideration; being open to audit is a close second.

The following briefly outlines a practical documented system structure but leaves the detailed style of the environmental manual, procedures, work instructions, etc. to the implementing organization. It is certainly a good idea to have a three- to four-level 'pyramid' type hierarchy of documentation, with the environmental policy at the top, which spreads sideways down through the manuals, procedures and supporting documentation. This is the structure to be found in the majority of organizations that have a management system and is deemed to be good practice.

Level 1: The environmental manual

Level 2: The procedures

Level 3: Work instructions/specific routines

Level 4: Forms, documents, plans, lists

This structure is illustrated in Figure 2.10. Each of these levels is described below.

Level 1 – The policy

This was discussed earlier in the chapter.

Level 2 – The environmental manual

This can be a rather brief document, which need only include the environmental policy and a broad description of how the organization has addressed the requirements of the Standard. As a document, with no commercially sensitive information, it can be sent to customers or other interested parties (at little cost) and is, in fact, a good marketing aid.

The opportunity should be taken to show how, for each clause of the Standard, the organization has procedures in place and decision-making processes. These should be referenced. This has the double benefit of showing interested parties (including external auditors) exactly how the Standard is addressed and, more importantly, making this clear for the organization itself.

Obviously, when compiling the environmental manual, if the organization cannot assign a procedure or a methodology of working to one of the clauses of the Standard then this must demonstrate a gap in the environmental management system.

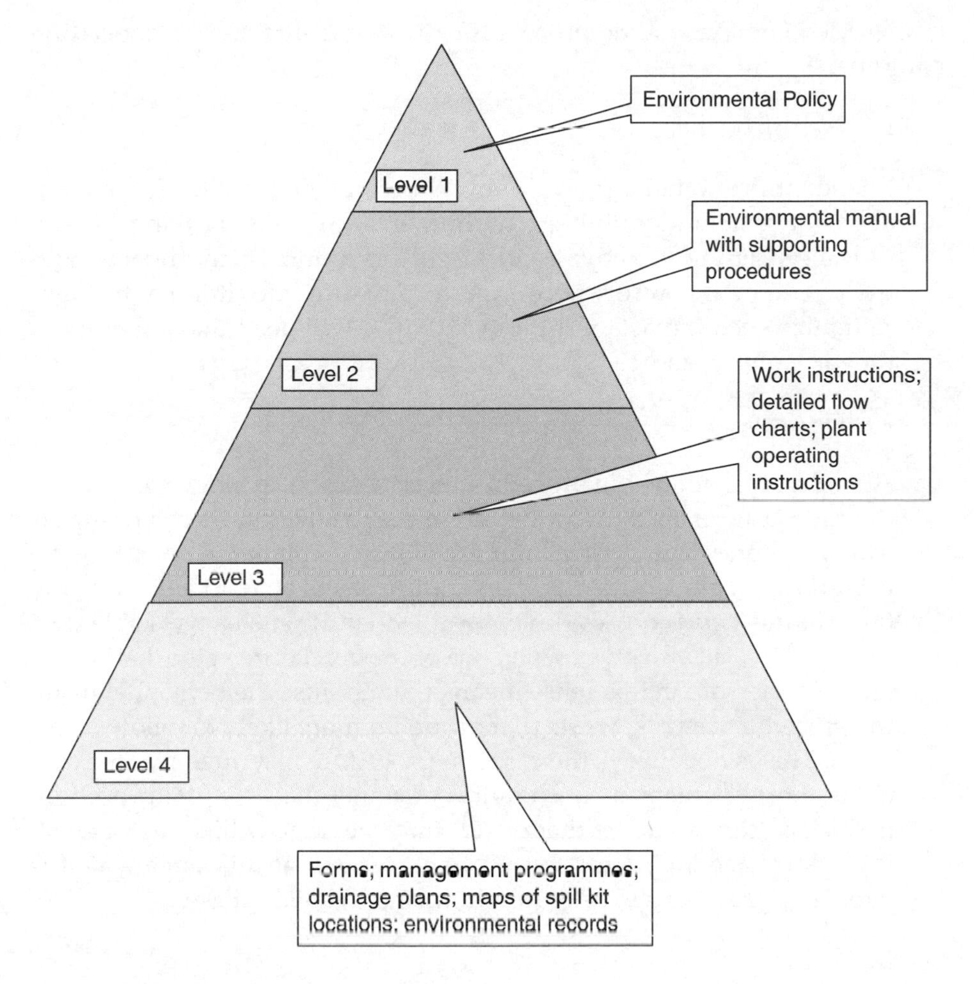

Figure 2.10 Hierarchy of documentation

Certainly, the concept of a 'signpost document' is a good analogy and this concept is encouraged by certification bodies.

Environmental management system procedures

The procedures will tend to dominate the bulk of the whole documented system. However, efforts should be made again to signpost, if possible, other existing systems. Thus, if there is a 'handbook' for operating a particular piece of equipment safely, perhaps with environmental considerations as well, then this handbook need only be referenced. This is preferable to writing a new procedure or instruction when the existing one is adequate. An outline process for the establishment of a procedure could be:

i) Identification:

The identification of a procedure will probably evolve from the management system identifying environmental impacts (ranking for significance, setting objectives and targets to reduce such impacts) and the operational controls necessary to measure, monitor, control and minimize such impacts. All these activities will need documenting in the form of procedures.

ii) Drafting:

An individual, following procedure identification, actually has to write the procedure. This is not a daunting task provided there are inputs from personnel who are involved (or will be involved) in operating the procedure. In practical terms, a procedure has to be workable and only those at the 'hands-on' level will be able to write such a workable document. This will also give ownership of the procedure at the 'hands-on' level. Ownership in this sense means that because the personnel using the procedure actually wrote it, they are far more likely to follow it correctly. This is not always the case if a procedure is written by an external author – perhaps a consultant. Personnel may feel the procedure has been 'thrust' upon them and may be less willing to obey it. Procedures can be in note form – hand-written at this stage – as it is likely that changes will be required during the pilot phase.

iii) Piloting:

Experience shows that it is extremely difficult for an individual to write down their daily tasks; they are of such a second nature that it is easy to make omissions, or to put the steps of a process in the wrong order. Therefore, the procedure should be piloted by other personnel. There may be subtle interfaces or interactions with other personnel or other procedures that the original authors had not considered because they are too close to their own area of responsibility.

iv) Revision:

It is inevitable that revisions will be required. For example, simple errors may need correcting; gaps in lines of communication may have been discovered; a new perspective may have been added.

v) Implementation:

It is important that all those personnel using the procedure understand its purpose. Some time should be invested in training personnel – introducing the reason for the procedure should be the focus of such training.

vi) Auditing:

Last but by no means least, the procedure once written, piloted, revised and implemented needs to be verified as to its effectiveness. This can only be checked by an independent internal audit – and the rule of the '6 Ws' is good guidance to follow. When writing a procedure, if it addresses the following questions, then it is likely to be meaningful and robust:

1) **W**hat does the procedure require?
2) **W**hen does the procedure need to be followed?
3) **W**here do the activities take place?
4) **W**hom does the procedure apply to?
5) **W**hy is the procedure carried out?
6) Ho**W** are the activities carried out?

Finally, although procedures have been discussed as being 'written', process charts or flow charts can be used instead. (Chapter 4, Figures 4.5 to 4.8 illustrate some simple process charts.)

Level 3 – Environmental work instructions/specific routines

It is not mandatory for the organization to include environmental work instructions in its documentation hierarchy if it is not appropriate to the business. However, in an organization with many processes, comprising of levels of alternative routes or recipes, there may be a need for separate, more specific and detailed 'routines' for operators. Otherwise, they may find it difficult to locate the relevant section contained within just one long procedure. This could lead to an environmental incident. Therefore, a work instruction at the point of the operation is a good idea. A note above a drainage sink, to remind operators that locally produced acidic waste liquids must not be poured down the sink, is a type of work instruction.

If work instructions become obsolete, and they are followed by an operator, then an environmental incident might occur. Obviously such instructions must be controlled as to revision status and location on the site. (See Section 4.4.5 *Control of Documents*.)

Level 4 – Forms and documents

Every organization uses forms be they hard copy or electronic. Forms allow personnel to record information in a structured manner so that other personnel can read and use this information. Forms also remove the requirement for the individual to remember every piece of information given. This means that less errors will occur in the management system as the system is not dependent on the frailty of the human memory. Well-designed forms can act as a prompt for individuals to record the correct quantity of information as well as the quality (that is, the usefulness of that information). Therefore, it makes sense to control the information demanded by the form so that nothing important is overlooked. Forms may be revised many times throughout their life to cope with changing circumstances. Such forms should be controlled, so that if a form is revised this should be indicated on it (Revision 1, 2, etc.).

Of course, the organization should review the 'significance' of the information on the forms and if it is deemed to be trivial, then there is no need to control such a form. It is good practice for any organization to review all the forms it uses, make a list and analyse what function each form performs. Most organizations who perform this task actually end up with less forms in circulation. The remaining forms are more focused in their purpose.

The reference to 'documents' covers all other 'reading matter' that the organization refers to in its environmental operations and decision-making processes. This could include copies of international standards, machine operation manuals, codes of practice, tables of calculations, etc. Again, the organization should review which documents are significant (in terms of the information they contain) and, where appropriate, ensure that they are given a controlled status. These manuals, procedures, work instructions and appropriate forms must ultimately be issued to personnel in a controlled fashion and this is addressed by the next sub-clause.

Clause 4.4.5: Control of Documents

The Standard calls for the organization to establish and maintain procedures for controlling all documents required by ISO 14001 to ensure that:

- They are legible and readily identified.
- They are periodically reviewed and approved for adequacy.

- The relevant versions are available at point of use.
- Obsolete documents are removed or if retained, suitably identified.
- Any documents of external origin are suitably identified.

Thus, this sub-clause is in effect stating that the EMS documentation, as required by clause 4.4.4, should be controlled or managed. The control of documentation can be exercised at the different levels of documentation (as in the three- to four-level pyramid hierarchy suggested in the previous section discussing clause 4.4.4). For example, stamps can be used to signify 'controlled' or 'uncontrolled'; coloured paper can be used, with or without a special logo, to make it obvious if an 'illegal' photocopy is being used.

Whatever method is used, it should always focus on the principle that the desired and planned information is available to those personnel that require it. There should never be any danger that out-of-date information, or the wrong data, is read or used by those individuals. If information is not important, or it does not matter whether it is up to date or not, then there is no need to control it. Marking it 'uncontrolled – not subject to update' – is a good mechanism to ensure that the casual reader can use that particular document for background information but is also prompted to seek assurance as to its current validity (for example, if a particular decision needs to be made based upon the information in the document).

Controlling documentation does give personnel confidence in their activities; they are sure that the decision they have made is the correct one. Indecision based upon mistrust of information received is responsible for many environmental incidents.

Signatures of the management representative (or even the Managing Director) on each page are also examples of control being exercised in documentation distribution. Consideration should also be given as to the cost-effectiveness of controlling documents. Does the time and effort spent controlling documents far outweigh any possible problems should there be a loss of document control?

Clause 4.4.6: Operational control

The Standard requires identification of operations and activities that are associated with the identified significant environmental aspects of the organization – in line with the environmental policy, its objectives and targets.

In effect, procedures are required to control and verify all functions, activities and processes which have or could have, if uncontrolled, a significant impact (direct or indirect) on the environment. Environmental impacts of the organization's suppliers come under the controls exercised under this sub-clause, as do those of contractors coming on-site or contractors used by the implementing organization. This is especially so if the suppliers' or contractors' methods of working are known to conflict with the organizations' environmental policy. Contractors may not be aware of the environmental consequences of certain actions (for example, dumping their waste into skips without segregation, or operating noisy drills in the evening when the organization may in fact have come to some arrangement with their neighbours not to do so). Many organizations have induction sessions for contractors as a matter of course (for example, on health and safety issues) so environmental awareness could be included in this existing mechanism.

However, the principle of significance should be considered when deciding what controls to put in place. For example, the contractor who comes on-site to clean office windows should receive less attention than the contractor who is laying new cables or drainage pipes on-site.

Organizations with existing ISO 9000 quality assurance 'process control' procedures should avoid merely relabeling some of these process control procedures as 'operational control'.

Unfortunately, some organizations have mistakenly seen operational control as an extension to the quality assurance system and have therefore ended up with very descriptive procedures which are unwieldy to use and maintain. At the early stages of systems implementation, such operational controls should only be in place to control the significant impacts. Other operational controls may need to be written later, as their level of significance increases through the continuous improvement process inherent in the system.

Similarities will be apparent in the structure of such procedures, compared to quality assurance systems (for example the 6 Ws as previously described). However, a different philosophical stance is required and instead of operating, testing and controlling for quality specification and tolerances, some rethinking is needed. For example, plant should be operated in a way that ensures that energy usage and pollution are minimized. This may require a more detailed procedure for start-up – when pollution, in the form of airborne particulates, may be greater until the process has stabilized.

Examples of operational control procedures

Waste management

Procedures should state who is responsible for ensuring legal compliance when waste is removed from site. The procedure should reference how the appropriate documentation is completed correctly and filed away for the prescribed period of time. Locations for the storage of waste should be described and instructions as to how segregation of waste is to be carried out described.

Procedures should detail further whether certain waste skips need to be protected from high winds and rain – which could blow or dissolve and wash some waste materials down surface water drains. Inspections of waste skips on site should be inspected for rust. Such deterioration could decrease their load bearing capacity or integrity, which could allow waste materials to fall onto public roads when being transported to the landfill site.

Best practice is to also conduct audits of waste carriers and landfill sites. This can range from a desktop review of waste carrier and landfill licences, to a physical audit of tracing an actual waste removal from site to the landfill site. Such an audit trail would also involve checking records and documentation, such as waste transfer notes. Landfill site licences generally are very specific in describing what is accepted at each site, and detail requirements for generating and maintaining records of waste.

This can be further extended to inspection of the waste carriers transport, i.e. lorries and wagons for roadworthiness and whether the driver has a spillage kit and is trained to use it in the event of a spillage onto a public road.

These are all best practical steps to prevent pollution and also avoid adverse publicity from the media and unwanted attention from the regulatory body.

Packaging waste

The European Directive on packaging waste places requirements on companies to recover certain percentages of waste and recycle set amounts. Companies can administer this themselves or join a compliance scheme. A procedure for the operational control of this waste should detail the responsible person who collects the data on packaging waste. Additionally, details of the prescribed time intervals for submission of this data to the regulatory authority should be included.

Contractors coming on site

The procedure should detail the requirements for contractors working on the site. A good practice is to show a brief video presentation of the organizations environmental procedures, followed by a test of understanding. Additionally, information packs should be issued containing for example:

- Site contact names and internal phone numbers (for environmental issues)
- What to do in an emergency
- list of actions which would be unacceptable such as disposal of waste into the site drainage system, burning of waste and not leaving the engine of vehicles running unnecessarily

Sub-contractors should sign a document as evidence they have understood the above.

Suppliers

A procedure for the purchasing department could include a list of the environmental responsibility criteria required of the organization's suppliers. Meeting such criteria would be a part of purchasing negotiations between the organization and the supplier. The objective would be to stipulate grades of material that have low environmental impacts in production, use and disposal.

Of course, the organization cannot merely use environmental criteria and nothing else. As stated earlier, the purpose of the Standard is not to prevent a business being successful. The commercial activities of the organization are paramount and must always come first in any decision-making event. If the business closes down then there is no environmental management system! For example, the procedure could detail that all existing suppliers be sent a questionnaire about their environmental probity as a first objective to obtain initial information. The procedure would then be written along the lines previously described.

Bunding inspection

Although bunding is built around storage vessels with the purpose of preventing leakage of the contents and, therefore, avoiding an environmental incident, in reality it may create a false sense of security. Maintenance may not be carried out because of this misplaced sense of security. The operational control procedure should be in the form of an inspection programme, and details of what action to take in the event of any damage to a bund wall.

Pipework inspection

Procedures for the inspection and maintenance of pipework and especially pipe joints should be implemented. It would be beneficial to identify which sections of pipe joints etc. could have the most significant environmental impact in the event of a failure and programme in extra inspections in these areas.

Monitoring for compliance with site licences

Although the regulatory authorities may periodically sample emissions or effluent, they would probably only issue a report or a concern in the case of non-compliance. Near misses or trends would not be a part of their reporting structure. Clearly to only rely on such reports creates environmental risks for the organization and it would therefore be best practice to have an operational control in place to periodically measure the parameters and note if there are any adverse trends emerging.

Future/Planned work

The Standard requires that any planned or new developments are considered and planned in advance so that the environment will not be compromised. Thus procedures need to be in place to give guidance in the instances of:

Construction work: Procedures should detail the forward planning and investigations necessary prior to any construction work on site. For intended excavations, operational controls should detail the responsibility for carrying out surveys not only for buried pipelines and any contaminated land, but also any ancient monuments or historic sites. The procedure should detail contacts and experts who could assist in identifying any protected species, trees and hedgerows, SSSI sites, marine habitats or controlled waters. This could include the names of agencies to be contacted if wildlife habitats are to be disturbed and any local liaison groups if the construction is in a built-up or residential neighbourhood.

Demolition work: Procedures should include some of the above considerations and if old buildings are the subject of demolition, then an asbestos survey should be carried out first. Such operational controls should also detail the steps to take in the event of discovering asbestos and the appropriate abnormal or emergency plans.

Housekeeping
In the context of ISO 14001, housekeeping is taken to mean 'the visual appearance of the site'.

Procedures should focus on defining locations for movable items, defining locations for redundant equipment, where tools should be stored, locations for discarded items and waste, etc. Housekeeping inspections should be detailed as to frequency and measures of compliance. Measuring the compliance of housekeeping can be subjective, and best practice involves the use of photographs to define the level of housekeeping required, area by area. With the advent of digital photography this method of measurement is becoming widespread.

Given that poor levels of housekeeping can contribute to the cause of accidents in the workplace, then some effort should be put into developing robust procedures. The certification bodies give this aspect of ISO 14001 what may seem too much attention but it is a visible sign of environmental improvement and sends a positive message to all stakeholders.

Clause 4.4.7: Emergency preparedness and response

The Standard requires that three components be addressed by the organization:

1 Establish and maintain procedures to identify the potential for, and the response to, accidents and emergency situations, in order to prevent and mitigate the environmental impacts that may be associated with them.

2 Review and revise the importance of learning from incidents. Obviously corrective actions will be taken and results of audits will be considered after the occurrence of accidents or emergencies (or even 'near misses').

3 Testing of emergency plans should be planned and the Standard indicates that periodic testing of such procedures should be carried out where practicable.

These three components are described below.

Establish and maintain procedures

An organization is well advised to draw up an 'Emergency Plan' or 'Crisis Plan' and to consider different levels of disaster – including the worst-case scenario. The worst-time scenario should also be evaluated which can occur at times of shift change, weekends and holiday periods when staff may be reduced.

The emphasis of such a plan will be placed upon ensuring that the business survives – that there will be the minimum of disruption of service, or supplies of product, to customers. Safety of individuals, staff and others will naturally be paramount but of course the environmental impact must be considered to address the Standard. The two factors (safety and environmental impact) can of course be interrelated – large volumes of toxic gases will cause environmental damage whilst posing a safety hazard at the same time.

A suitable plan could cover:

- Identifying critical assets – usually major items of plant
- Defining the disaster team and describing responsibilities
- Emergency services – how to contact them
- Enforcement agencies – how to contact them
- Listing communications – including out of hours phone numbers
- Holding regular exercises – to test the system
- Setting up a media response team – details of who is responsible for contacting the media

Examining the function of the media response team further is worthwhile. It is tempting in the event of an environmental incident to offer no comment to the press. However, other parties, who may be hostile to the organization, will comment. This could be disgruntled neighbours, competitors and others who will take the opportunity to repeat rumours or provide media with exaggerated accounts of the incident.

The media will find out, despite efforts to deter them. An attempt to hide issues will usually be uncovered later and will ultimately reduce the credibility of future communications. If, for example, there is a large spillage to a local stream or river with a high amenity value and the organization does not control the media input, the information any interested party receives may be inaccurate at best and hostile at worst. This could influence future relationships with the media. The news-gathering media work to strict deadlines and information is required from whatever source. It is best, therefore, that this source is the organization itself.

Possible emergencies will most certainly have been identified during the preparatory environmental review and suitable responses formulated. At the simplest level of response, this may include a list of competent personnel who can be contacted (with alternatives) in the event of an emergency situation. Provision should also be made for off-site availability of the information needed to contain the disaster in case access to the site is denied on the grounds of safety. The main objectives of the fire service are to control the fire and/or chemical spillage and save lives.

They may cordon off the site and prevent access to staff who require telephone numbers (to make contact with other personnel), access to records or emergency procedures, etc. It is worthwhile noting that even if a fire situation is handled correctly by the fire service, environmental problems can be caused by contaminated water used by the fire services. Such water, contaminated by combustion products, may enter canals, streams and drains, and emergency plans should take this into consideration.

It is also important to remember that just because risks are low, it does not mean that emergency plans are unnecessary. Without an emergency plan, minor incidents can escalate into major ones.

Review and revise the importance of learning from incidents
In the case of near misses it is important that such potential incidents are recorded and reviewed and not hidden away or merely forgotten. Such incidents indicate areas of risk which on other occasions may turn into environmental accidents.

Testing of emergency plans
There may be situations where full-scale testing is not practical and thus consideration should be given to desk-top exercises that can be played out. Examples of such testing include:

- Can key individuals be contacted in an out-of-hours situation?

- If staff are injured can relatives be contacted?

- Are keys to certain areas (for example, solvent stores) available out of hours?
- Are vacations or absence due to sickness covered by alternative personnel?
- Can emergency services access the site – day or night?
- If there is ice on the road during winter would this prevent heavy vehicle access – fire-fighting equipment for example?
- Do the organization's fire hydrants function correctly? Are they maintained and tested?
- Does the fitment on the hydrant fit the fire services hoses?
- Toxic gases may be released during a fire. If so, what are these gases? In which direction is the prevailing wind? What is the likely area that will need to be evacuated for safety as well as environmental reasons? What information will be given to the local police in such an event?

Test emergency scenarios at the shift change from day to night shift for example. During such shift changes there is an overlap as incoming staff come in early so that the outgoing staff may brief them as to events that have happened etc. Thus, for a short period of time, there could be double the number of people on site, causing extra congestion of roads, and lines of command may become blurred during this period.

Some organizations ban the use of cellular phones during normal operations because of safety issues. Would staff be able to use them in an emergency? If so, this should be written into the emergency procedure.

Some organizations have wind direction indicators on site so that in the event of an emergency (fire or noxious fumes escaping), staff can move away in the safest direction. Test scenarios should ascertain whether such indicators can be seen visually from all points of the site.

Some organizations have an incident centre that would reasonably be expected not to be influenced by a disaster – i.e. underground or remote from flammable areas. Test scenarios should check whether this isolation is still valid.

The time to put an emergency plan to the acid test and seek the answers to the above is not the day of the real event. Therefore, an organization should evaluate its risks of environmental emergencies and evaluate the extent to which it should carry out testing of its procedures.

Clause 4.5: Checking and corrective action

As with any human undertaking, errors will occur and systems need to be in place to check that events are happening as they should and that, if errors do occur, suitable corrective actions are taken. The following sub-clauses of the Standard require such activities to be carried out in a structured fashion to ensure an effective EMS.

Clause 4.5.1: Monitoring and measurement

This requirement of the Standard is for an organization to establish and maintain procedures to monitor and measure (on a regular basis) the key characteristics of its operations and activities that can have a significant impact on the environment.

It is important to understand the differences between monitoring and measurement to comply fully with the Standard.

Monitoring generally means operating processes that can check whether something is happening as intended or planned. In some respects auditing processes address this, but also operational control procedures will apply. Thus if an operational control states that housekeeping audits will occur twice weekly then this is a monitoring process, i.e. the site is checked weekly for 'good housekeeping practices'. This could also involve 'visual' checking of the integrity of bunding around solvent storage tanks for example.

Measurement tends to mean that the size or magnitude of an event is measured, calculated or estimated with a numerical value assigned.

This could include procedures for weighing wastes sent to landfill; amount of gas or electricity consumed per week, measuring noise levels at the site boundary etc. Additionally, any equipment used to calculate or estimate such numbers should be suitably calibrated so that a high level of confidence is gained that the numbers are indeed a true representation of the facts.

Clause 4.5.2: Evaluation of compliance

This is a requirement which links to clause 4.3.2 *Legal and other requirements* to ensure that breaches of such applicable environmental legal requirements do not occur. This clause is best addressed by having robust procedures in place to periodically review the level of compliance to those operational controls, or work instructions, which specify activities to ensure legal compliance.

Thus, as an example, a work instruction may detail that samples of effluent are checked daily in the site laboratory to check that heavy metal concentrations are not progressively rising, which could lead to a breach of the consent parameters, hence a breach of legislation.

By auditing such an instruction and reviewing results obtained, then an evaluation can be performed. Audit checklists should indicate clearly that compliance was verified. If during this audit an area is identified where there is not full compliance with legislation, then corrective and preventive actions should be identified and executed to ensure a return to full compliance.

Clause 4.5.3: Non-conformity, corrective and preventive actions

The Standard requires that the organization shall establish and maintain procedures for controlling non-conformities and for taking corrective and preventive actions to mitigate any impacts caused.

Thus the organization must have the capability of detecting non-conformances and then setting up mechanisms for correcting each non-conformance. Further, it should be able to put into place systems that will prevent a recurrence of the same non-conformance.

A moment should be taken here to consider what is understood by 'non-conformance' because not every organization knows what an environmental non-conformance is. Some time should be spent defining the term 'environmental non-conformance' to all interested parties.

For example, not following a procedure is an easy non-conformance to identify. However, this is really a management system non-conformance. A true environmental non-conformance could be a cracked bund wall that has not been repaired as programmed – thereby increasing the risk of an environmental incident.

The requirement of the Standard goes on to state that any corrective or preventive action that is taken shall be appropriate to the magnitude of the problem and commensurate with the environmental impact encountered. This is to ensure that the organization is mindful of that word 'significance'. Referring to the cracked bund wall above, suppose that it is designed to contain only process waters of low toxicity. Clearly, a similar cracked bund wall intended to contain a cocktail of waste solvents has a much higher priority on the corrective action agenda. Higher priority must also be given to monitoring and maintenance costs than the former bund wall. Therefore the organization is using its finite financial resources in the most effective way.

The closing paragraph of this sub-clause merely calls for an organization to record what corrective actions were taken and to make revisions to any relevant documented procedures. Again, using the above example, a possible change to the maintenance procedure would be to increase the frequency of inspection of the higher-risk bunding.

Clause 4.5.4: Records

This sub-clause states the organization shall establish and maintain procedures for the identification, storage, protection, retrieval, retention and disposition of environmental records.

It is as well to remember that a record of an activity is invariably required some time after the event has occurred; when the person who created or added to the record has left the organization and that everyone connected with the 'incident' or 'occurrence' cannot remember what happened.

These records shall include training records and the results of audits and reviews.

Records need to be kept by any organization to demonstrate that previous activities have been in compliance with legislation, including:

- Discharge consents to sewer
- Process authorizations and variation notices
- Controlled waste consignment notes
- Special waste transfer notes

Other necessary records include: accidents, incidents, training, calibration, performance monitoring, internal audits, training, management reviews.

For example, due to packaging waste regulations in the UK, records such as weighbridge tickets will have to be kept as the organization needs to demonstrate to the Environment Agency that recycling and recovery of waste are meeting legislative requirements. The transfer note system requires that both parties keep a copy of the transfer notes and the description of the waste for a minimum period. There may be an occasion when the organization has to prove in a court of law where the waste came from and what they did with it. A copy of the transfer note may also need to be made available to the enforcement authority if they wish to see it.

The Standard also requires that records are legible, easily identified, easily retrievable and protected from damage or loss. The minimum length of time that they are kept should be documented. The approach taken by most organizations is to review what records they need to keep (not all records need to be kept but any record that has an environmental implication should be reviewed as to whether it needs to be retained); where they are kept; and for how long. A procedure can consist merely of a list detailing the above; this has the advantage of being easy to maintain and audit both by internal and external auditors.

Records, of course, also refer to electronic storage. A means of protecting such information through regular back-ups and storage off-site must be established. As some of the information should be 'readily available', elaborate password systems must take this into account by allowing access to those who are authorized.

Clause 4.5.5: Internal audit

The Standard requires that an organization carry out periodic planned environmental management system audits in order to:

a) Determine whether the environmental management system conforms to planned arrangements (controlling and minimizing the significant environmental impacts) and meets the requirements of the Standard.

b) Provide feedback to management of the results of such audits. Such auditing should be performed on a planned and scheduled basis to reflect the environmental significance of the activities being audited.

Audit methodology
Considering part a) of the sub-clause there are two components to this:

- Auditing whether the EMS is delivering environmental improvements, i.e. 'Performance' auditing (see Figure 2.11).

Performance auditing is the process of verifying that the environmental management system is delivering improvement in performance. That is whether environmental objectives and targets are being met.

- Compliance auditing.

Compliance auditing is more of the style of 'traditional' auditing, i.e. ensuring that procedures are being followed in order to comply with the requirements of ISO 14001.

This type of audit ensures that personnel are following procedures: i.e. taking measurements when and where they should; reviewing and updating the legislation register; generating the appropriate records, etc. Such auditing methodology will be second nature to an organization that already has a documented quality assurance system meeting the requirements of ISO 9001.

Some organizations keep the two audits very distinct – performed at different occasions and by a different set of auditors. Others use the quality assurance auditors for compliance auditing and only use their specifically trained environmental auditors for the performance auditing. Still other organizations use a combined approach with a team of internal auditors or issue their auditors with basic 'environmental checklists' compiled by an environmental specialist. There are no rights or wrongs here – it is whatever suits the culture of the organization best.

Some organizations carry out compliance audits only but have additional mechanisms to monitor their environmental management programme. Some have regular monthly meetings where each assigned project manager reports on progress or otherwise, of environmental improvement programmes. Smaller organizations may have a monthly review of target success and reporting back to senior management. These review processes, although not strictly classified as internal audits, are in fact just that. An internal scrutiny of planned events. The organization must choose which is the best method to use to check on their progress and in the final analysis,

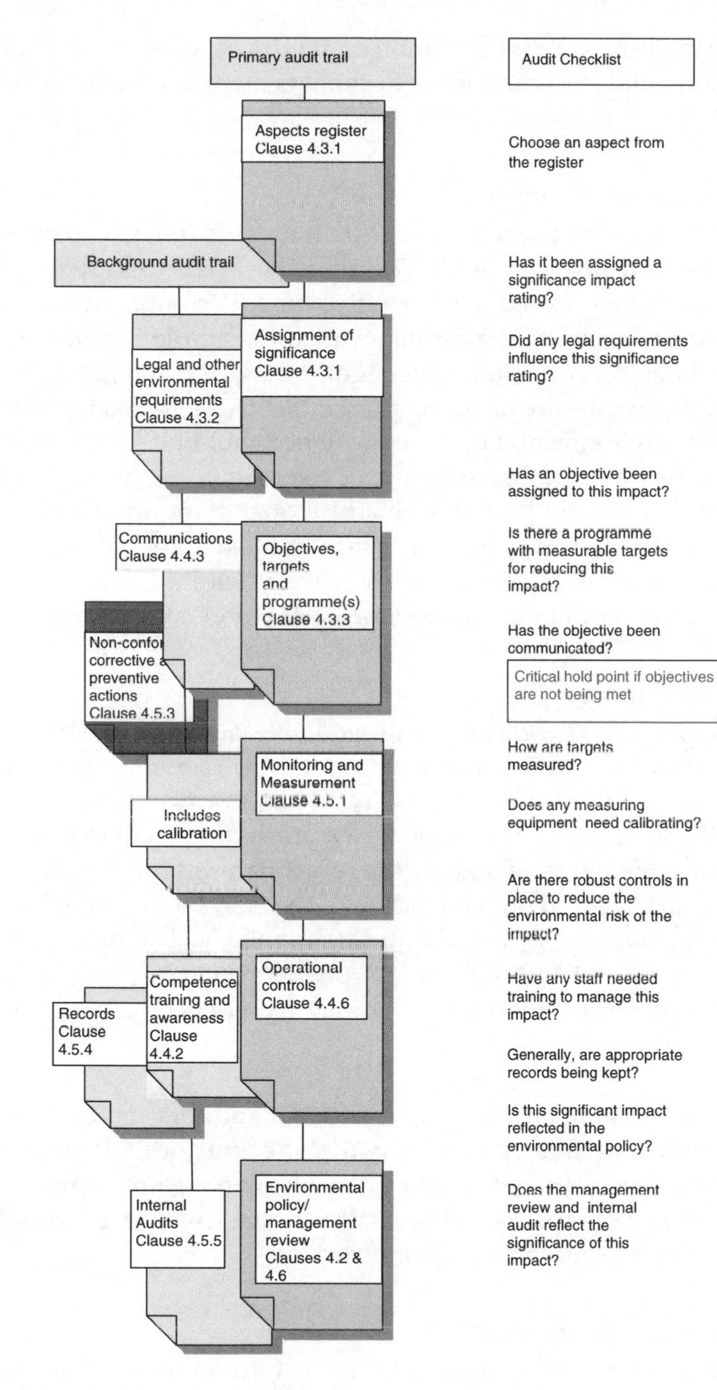

Figure 2.11 Performance auditing

neither one method is superior to another. It is the level of scrutiny and corrective actions and reporting back to management mechanisms that are paramount.

Environmental non-conformance

It is important that the internal auditors can identify 'environmental' non-conformances during the audit (as distinct from 'compliance' non-conformance which would be identified as in 'compliance' auditing, above). A compliance non-conformance is, for example, when a member of personnel neglects (due to an oversight) to log onto a register the daily amounts of one waste stream going to landfill. The action is taken but not recorded. An environmental non-conformance could be a measurable target (such as the monthly figure for use of recycled paper not increasing as planned). If this has not been identified by operatives and management, for whatever reason, and corrective action not taken, then this may affect the objectives set. It may jeopardize a statement within the environmental policy and constitute a potential environmental non-compliance.

Auditor qualifications

Some organizations use existing and available quality systems auditors from an existing ISO 9001 system for the compliance auditing (as outlined above). There should be no difficulty with this practice as these auditors will be well able to audit for compliance against documented procedures. Qualifications for effectiveness audits are different. The auditor would need to have a grasp and understanding of the Standard and the EMS, and a broad understanding of environmental issues. Such requirements can certainly be achieved through an independent learning process with a combination of formal training and direct experience (perhaps initially under the guidance of consultants).

The Annex to the Standard also points out that auditors should be reasonably independent of the area or activity that is being audited. Again, this is only common sense. An auditor, auditing his own area of competence, is hardly likely to be impartial – especially if faced with a potential non-conformity directly traceable to their error!

Reporting back to management

As noted initially, the Standard calls for some form of feedback to management on the results of the audits. In truth, this is common sense, because if the results of the audits demonstrate major discrepancies in what was planned (through objectives and targets) and what is actually being achieved, then management needs to reconsider the effectiveness of the whole system in order to fulfil its obligations as set out in the environmental policy.

Auditing procedure

The methodology for performing the audits should be established within written procedures. How else can an internal auditor know how to conduct the audit? Frequencies of auditing should be specified – written in a schedule, plan or even a chart – and this should take into consideration the results of previous audits. Many non-conformities raised at the last audit should trigger off a more frequent re-audit until it is established that the corrective and preventive measures have worked. Finally, other types of audit used by organizations include external site audits – housekeeping (litter etc.), visual impact from the outside, odours, dust. These additional audits can only add to the effectiveness of the compliance and performance audits.

Clause 4.6: Management review

This clause requires that the organization's top management shall, at planned intervals that it determines, review the environmental management system to ensure its continuing suitability, adequacy and effectiveness.

Again, common sense dictates that once a system is implemented, there should be a review process to test whether what was planned does happen in reality. There is no correct way to perform an environmental management review – it must suit the organization's culture and resources. As the Standard refers to 'top' management, this does indicate that a certain level of seniority of personnel should be present at such reviews, to demonstrate commitment. However, there are certain minimum areas to be reviewed and one option, used by most organizations, is to have a standard agenda for each meeting. The first point on the agenda should be a review of the Environmental Policy. This is the 'driver' for the whole system. Senior management should be able to examine it and say with confidence that what was planned (say 12 months ago) as stated in the policy, has occurred or that substantial progress has been made. Thus a typical agenda could be:

- Are the objectives stated in the Environmental Policy being met?
- Does the organization have the continuing capacity to identify environmental aspects?

- Does the system allow the organization to give a measure of significance to these aspects?
- Have the operational controls that were put in place achieved the desired levels of control?
- Are effective corrective actions taking place to ensure that where objectives are in danger of slippage, extra resources ensure a return to the planned time-scale?
- Are internal audits effective in identifying non-conformances?
- Is the environmental policy sufficiently robust for the forthcoming 12 months?

It could be argued that during the early months of the implementation period (perhaps prior to certification) these cyclical reviews are not appropriate and they should focus on just the progress of the implementation of the system. This is a reasonable viewpoint but, as the system approaches maturity, a review as above is beneficial at intervals of 6 to 12 months.

It would be prudent for the organization to perform one full management review, following the procedure, prior to the on-site audit, to demonstrate evidence of implementation to the certification body.

If it is concluded that the set objectives are being met, the organization is well on its way to minimizing its significant environmental impacts and thus complying with the requirements of the Standard.

Continuous improvement

Finally, the concept of continuous improvement is much misunderstood. An organization may have an objective to reduce its amount of waste to landfill. A target of 50% reduction may have been achieved over a period of some 3 years – a comendable environmental achievement. It could be, however, that to reduce this further (to demonstrate continuous improvement) would not be cost effective, perhaps the law of diminishing returns starts to operate at this level of reduction. The organization may well suffer financially to continue in this direction.

The intention of ISO 14001 is to recognize this and really wants the organization to then focus on a different area of improvement. Operational controls will be in place to maintain the reduced waste to landfill quantities. Management can look elsewhere in its processess for improvement.

Thus, over a long period of time, the organization can demonstrate improvements, year upon year, although of course the improvements will be in different areas or departments of the organization.

It is achieved by continually evaluating the environmental performance of the EMS against its environmental policies, objectives and targets for the purpose of identifying opportunities for improvement.

And, although we are talking about continuous improvement in terms of reduction of tangible waste or measurable and costed aspects (reduction of use of electricity, fuel, etc.), continuous improvement in the EMS is allowed. Some organizations have, for example, recognized shortfalls in measurements. Thus an environmental objective could be to improve the accuracy of the data they receive for electrical energy consumption. They may only have a site wide figure. The objective is to find methods to obtain energy consumption from each piece of plant. (They can then focus more easily on less efficient plant for replacement for example.) Such an environmental objective does in itself not reduce electricity consumption, but the system is improved allowing the possibility of reducing energy use in the future. Overall, the EMS will have been enhanced.

One organization may decide that, due to its processes, it cannot detect quickly enough if they are going out of consent with their effluent treatment, resulting in higher effluent treatment charges rather than prosecution from the authorities. An objective would be to investigate methods and or new detection equipment to prevent this occurring in the future.

Summary

The requirements of the Standard can seem daunting at first, but by obtaining commitment from top management, with methodological planning and a good understanding of the concepts of ISO 14001, the implementation need not be too difficult and should be well within the reach of the smaller as well as the larger organization.

However, not only does this achievement require management commitment during the implementation phase but, arguably, more commitment is required after the ISO 14001 certificate is obtained. The process of assessment and what follows after certification is described in the next chapter.

Chapter 3

The assessment process

Introduction

The accepted method of validating an organization claim to be ISO 14001 compliant is for them to be assessed and certified by an impartial, independent external body, an accredited certification body. Assessment and certification of an organization's EMS provides a balanced, impartial measurement and verification of:

- Compliance with environmental regulations
- Commitment to care of the environment
- Management of environmental risk

The award of a certificate to a successful organization demonstrates to all interested parties that the organization is committed to environmental responsibility.

In practical terms (and until a better method is discovered), an auditor, or a team of auditors, physically 'measures' the EMS against ISO 14001 requirements using the tools of questioning, interviews, discussions and by making reference to objective evidence, such as documents and records. After weighing up all the evidence, the auditor will make a judgement as to whether the system satisfies all clauses of the Standard. This assessment process has been adapted from the more mature discipline of quality assurance certification, ISO 9001, which has been operating worldwide for many years.

Certification bodies themselves have their conduct audited by 'accreditation bodies' (see Appendices I and II for details) who are, in turn, subject to rules, regulations and controls operated by a higher authority (usually an arm of a national government). These controls exist to ensure that the certification body demonstrates impartiality, the correct methodology, and an adequate level of competence in the conduct of the certification process. Such a certification process can then withstand scrutiny from all interested parties and has international credibility.

The regulatory framework for certification bodies (see Figure 3.1)

In the UK, for example, UKAS (United Kingdom Accreditation Service) is the higher regulatory body governing the activities of those certification bodies offering accredited certification to ISO 14001. Their experience in such matters has been gained through quality assurance, ISO 9001, which was valuable in setting accreditation criteria to enable accredited ISO 14001 certification.

Harmonization of accreditation criteria – that is, an agreement on accreditation processes – has occurred world-wide under the IAF (International Accreditation Forum, see Appendix II).

The considerable effort that has gone into achieving global harmonization is a reflection of the concern that ISO 14001 be interpreted as uniformly as possible, thereby ensuring uniformity of auditing from country to country, continent to continent.

Accreditation Criteria (see Appendix II for summary points) is designed to guide certification bodies towards uniformity of approach. Although

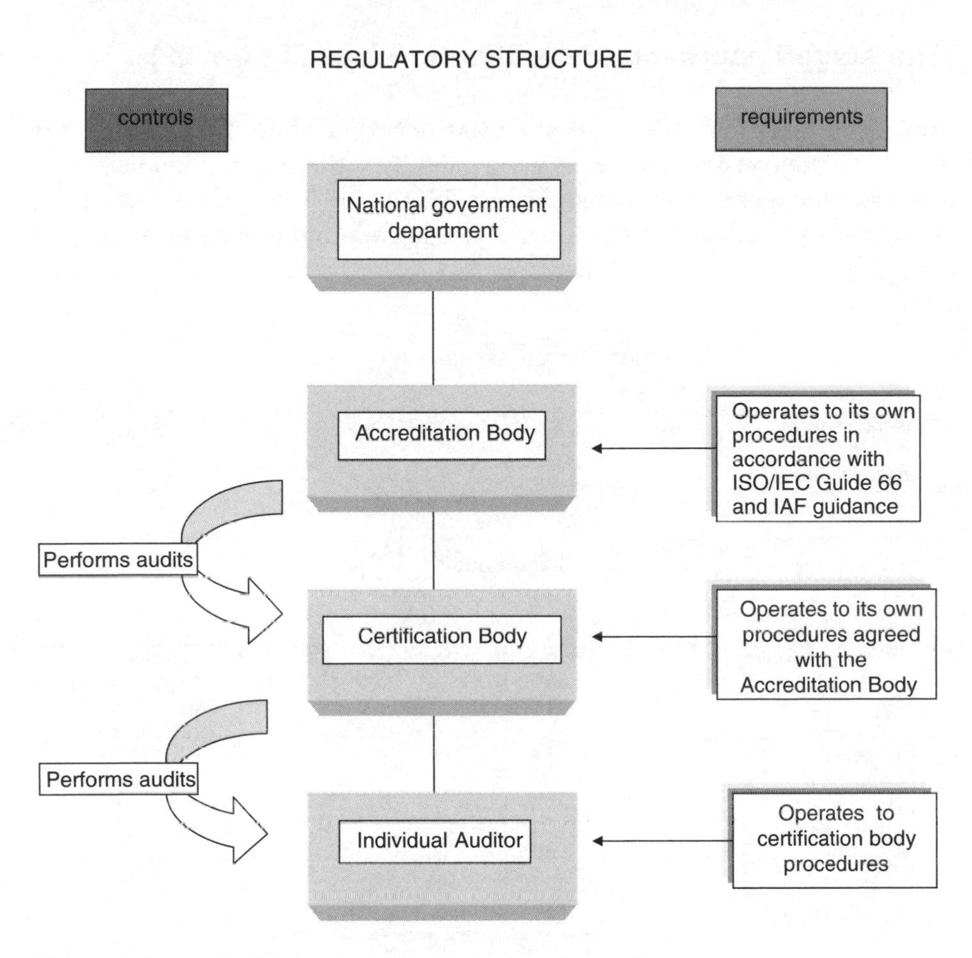

Figure 3.1 Regulatory structure

designed for certification bodies, an implementing organization would do well to examine the full document as it will help to provide an insight into what the certification bodies will be looking for when establishing the conformity of an EMS with the Standard. The document describes the staged approach required, as well as the certification bodies' management structure, auditor training and competence. Further it states that a minimum 'two stage' approach should be taken unless there is compelling justification for not doing so.

It can be considered to be an 'overlay' specification on top of ISO 14001, which is mandatory for auditors and certification bodies alike, and whose function is to ensure consistency of auditing.

The staged approach to certification (see Figure 3.2)

This staged approach to certification has not come about by accident. It is based on the considerable experience gained by the certification bodies of quality management systems such as ISO 9001. This process has been designed to facilitate certification and has been found to operate extremely

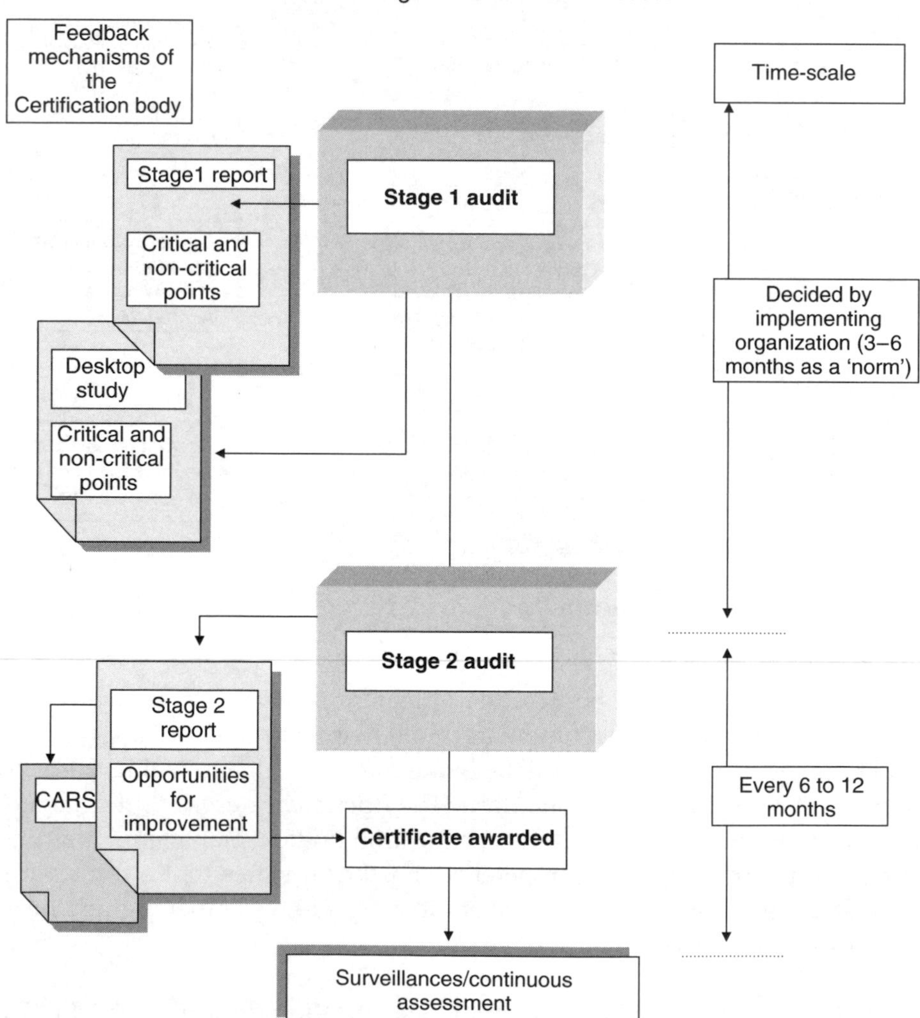

Figure 3.2 ISO 14001 staged assessment process

well for implementing organizations. The rationale is that at each stage the organization has the opportunity to rectify problems and address omissions prior to the next stage. For example, it would be in no one's interests if a team of auditors arrived on-site to perform the stage 2 audit and, within a short space of time, found so many omissions and errors in the system that the audit had to be aborted. While this demonstrates that certification is not achieved without some effort, it also wastes everyone's time, effort and money and can demoralize the organization's staff who will have put much individual effort into implementing the system.

The stage 1 audit

The objective of this stage of the assessment process is five-fold and is to:

1) Ensure that the EMS is based soundly upon the identification of environmental aspects

The adequacy of the PER (where applicable) is verified and evidence that the EMS takes the findings of the PER into account is sought. Reassurance is sought that a sound and logical methodology is being used to:

- identify environmental aspects;
- determine a measure of significance;
- and apply the resources of the organization to address such significant environmental impacts.

Indeed, a large proportion of the audit time may be spent on interviews with key staff in order to test the decision-making processes. The auditor will also need to perform a physical tour of the site, both internal and external areas, in order to gain an overview of the site, its size, complexity, neighbours, processes, activities and so forth. From the above the auditor will compare his assessment of the significance of aspects with the organization assessment and clearly there should be a high level of agreement between the two.

2) Ensure that the internal audit methodology is correct

At this early phase, the external auditor needs to examine whether the internal audit process is conducted with the concept of environmental importance or significance in mind, and whether the process is capable of identifying and ensuring correction of environmental non-conformities.

If an organization has an existing ISO 9001 quality assurance system, the external auditor will be mindful of the fact that the concept of 'self-policing', or internal audits, may be within the organization's culture. However, the concept of performance auditing (see Chapter 2) and strong linkages between scheduling of audits and significant impacts, may not be fully developed. Auditors may well raise concerns if audit schedules, for example, follow the numerical order of the clauses of the standard to consecutive months: clause 4.1 to be audited in January; clause 4.2 to be audited in February, and so on, rather than schedules based upon significance.

However, an audit schedule 'guide' may be sufficient in the early phases of implementation until experience and knowledge of the organization's environmental impacts is gained.

Whatever the method used, the company must be able to explain to the certification body auditor the reasons and rationale for the method of scheduling employed.

3) Determine the organization's state of preparedness for the next stage of the assessment process

The auditor must gather evidence to demonstrate that the EMS is designed to achieve compliance with regulatory requirements and policy objectives: in particular, continual improvement of environmental performance. The auditor will look to see if implementation has commenced rather than full implementation throughout the organization at this stage.

Additionally, the auditor will need to review the documented system to ensure that it addresses all clauses of the Standard. This is called a 'desk-top study'. Clearly, if not all the requirements of the Standard have been addressed within the documentation, then certification cannot be allowed. As a note, some certification bodies perform the desk-top study during the stage 1 audit. Others treat it as an additional separate sub-stage between the stage 1 and the stage 2 audits and generate an additional report. Both approaches have merits. However, the desk-top study must be completed before the stage 2 audit.

The ease of performing this desk-top study varies. There is no set structure for the documentation and it will vary from organization to organization. However, the auditor must be able to find a corresponding reference somewhere in the organization's documentation for each clause and sub-clause of the Standard. Because the Standard is generic, an organization may interpret some of the clauses in an unexpected fashion. The auditor will look for the rationale behind this interpretation and, if not convinced, will raise a query on this interpretation in the stage 1 report. A signpost document – the environmental manual – enables the auditor to see how the organization has considered each point in the Standard. More importantly, it gives the organization the confidence that it has indeed covered all the necessary clauses!

The desk-top study can be performed at the auditor's place of work, or on the organization's site. The latter has the advantage of immediate clarification for the auditor of an unfamiliar set of manuals and procedures. Against this the auditor may feel he cannot do a professional job if the facilities offered by the organization do not match expectations (for example, adequate office facilities with minimum distractions).

The auditor may be presented with the most typical structure of documentation i.e. the 'pyramid, of policy, manual, procedures and work instructions and forms (see Chapter 2 section 4.4.4 *Documentation*). Such documentation should include a list of the environmental aspects of the organization and a list of the significant impacts. Any PER that is relevant could be included but a PER is not assessed by the auditor. However, if the auditor feels that the preparatory review is superficial and does not identify the real environmental issues, and the system is based upon this, then questions will be raised regarding the fundamental basis for the whole of the EMS.

The legislative framework documents should be reviewed including:

- Any relevant national laws
- Permits and authorizations for processes operated
- Copies of water and effluent permits (abstraction licenses, discharge consents)

Plus other background information which could include:

- Site plans – including topography
- Drainage plans
- Prevailing winds (for nuisance issues or fire emergency)

- Site boundaries
- Neighbours (domestic dwellings, farmland, amenity areas)
- Any local areas of protected species – wildlife for example
- Any aquifers
- Any canals, streams, rivers

4) Plan and allocate the resources for the next stage of the assessment process

Much of this, of course, depends on the outcome of the stage 1 audit findings, the number and severity of the findings dictating the time-scale for the next stage (see 'report' at the end of this section). The organization in many respects controls this date, as only they know the resources available to address identified areas of weakness before the assessment process can move forward.

The skills required of the certification assessment team will also need to be reviewed. This review would be prompted by the discovery during the stage 1 audit of more complex environmental aspects than was originally understood by the auditor. This could also arise if the organization has not fully understood its own environmental aspects and not documented them.

The auditor on the stage 1 audit may not possess adequate experience in all of these more complex or additional environmental aspects to perform a fair and valid judgement during the stage 2 audit. It would then be appropriate to discuss this with the organization and make arrangements to include a specialist, or technical expert with the appropriate expertise, on the stage 2 audit team.

The required number of days to perform the stage 2 audit is also estimated at this stage. This may have been indicated to the organization previously for budgetary purposes so this exercise is a confirmation, or otherwise, of the original estimate. The estimate itself is based on several factors, such as:

- the number of staff on site;
- the complexity of the processes and their resultant environmental impacts;

- the physical size of the site;
- geographical location. Travelling long distances will affect the estimate as certification bodies have to pay their auditors the same whether they are auditing or travelling.

The auditor must also consider the arrangements for future surveillance, or continuous assessment visits (see Surveillance later in this chapter).

5) Collect necessary information regarding the processes of the organization

Further understanding of the organization's processes will need to be obtained. This will include where applicable 'flowcharts' of the processes or explanations of the manufacturing technology or chemical reactions.

Conduct and progression of stage 1

This section gives more detail on the time management and the formalities of stage 1. All certification bodies will operate to broadly the same protocols but there may be minor differences. Such differences should be few and will not affect the integrity, purpose or outcome of each of the stages.

The opening meeting

At the opening meeting, it is expected that the organization's senior management will be present. This is the forum for the team leader to introduce the audit team to senior management (and vice versa) and to explain several points of the conduct of the audit. An itinerary will be drawn up or it may well have been sent and agreed in advance. The itinerary is discussed and if there are conflicts of timing etc. then these can be accommodated by the team leader. It is best to keep to the logical order of the itinerary. However, minor alterations to such things as lunch times, or conflicts with essential business activities, are understandable and can be worked around. The auditing team ensures that the organization's personnel are kept informed as to the audit's progress. A typical opening meeting would have on the agenda:

- *Confidentiality*: All auditors have signed confidentiality agreements with their employer (the certification body) stating that they will not divulge information (gained during the assessment) to any third party.

- *Explanation of the purpose of the stage 1 audit*: The stage 1 audit is not a pass or fail exercise. It is intended to bring any fundamental problems to the surface so as to pave the way for a smooth transition to stage 2 of the assessment.

Progression of the audit

The stage 1 will then occur along the lines detailed in the previous section, ensuring that all five of the objectives are addressed (see the stage 1 objectives above).

The auditors' knowledge of the organization will be increased so that a more appropriate, meaningful and relevant stage 2 audit can be carried out.

The closing meeting

At the conclusion of stage 1, the auditor or auditors will pool their findings and reach agreement as to whether to recommend that the stage 2 audit can go ahead unreservedly or that the organization has further implementation work to do. This is detailed in the report.

The report

The organization receives a report detailing critical and non-critical findings. The organization therefore has the opportunity to address any shortcomings as quickly as possible.

The report will detail critical and non-critical points for the organization to address. Critical points being those that unless addressed will prevent certification at the stage 2 audit. Non-critical points tend to be points for improvement, or development or points of clarification and may lead to minor corrective action requests at the stage 2 audit but are unlikely to prevent certification. Development points can also include recommendations of best practice to assist the organization.

In conclusion, the objective of the stage 1 audit is generally to ensure that the organization is in no doubt as to its state of readiness for the next stage of the certification process.

The stage 2 audit

The objective is to verify that the EMS has been implemented throughout the organization and that the organization is complying with its own policies and procedures, and achieving regulatory compliance and continual improvement of environmental performance.

Note that accreditation criteria is not very specific on the point as to the length of time required to demonstrate full implementation. This will vary from organization to organization and complexity of the system, number of employees and other factors. However, realistically, a period of about 3 months has been found, in the past, to be about the minimum length of time to demonstrate implementation. Thus the auditor will be looking for at least this time-period, of procedures being followed and records generated in the areas of:

- Identification and evaluation of environmental aspects
- Linkages between identified significant impacts and objectives being set
- Evidence to show that targets are being monitored, measured and reviewed as to progress
- Audits being carried out to a plan or schedule focusing on the most significant impacts first
- Checking and corrective actions taking place
- The environmental policy reflecting the site's activities
- Operational controls in place robust enough to prevent breaches of legislation

Conduct and progression of stage 2

The stage 2 audit has some processes that are very similar to those of the stage 1 (for example, opening and closing meetings) and these will not be repeated in detail. However, there are some differences, and these will be focused upon. The stage 2 audit is a more structured audit, and compliance with documented procedures is being sought.

Prior to the stage 2 audit, the team leader will send to the organization an itinerary showing the intended auditing activities. However, the itinerary will be structured in a way that allows the auditor to audit the requirements of the Standard by following 'linkages' in the system. Only minor amendments will be allowed (to fit in with the organization's start and finish times, for example). The itinerary will again refer to an opening meeting at the start of the audit and a closing meeting at the end of the audit.

The opening meeting

As at the stage 1, the opening meeting will be a forum for the team leader to explain how the stage 2 audit is conducted and how it differs from the stage 1. Additional team members, such as a technical expert (referred to in the stage 1 objectives), will be introduced, if necessary. The itinerary is discussed and, as at the stage 1, minor alterations can be accommodated by the team leader although it is best to keep to the logical order of the itinerary. The overall objective is to ensure that the organization is kept informed of the audit's progress by the audit team. Areas of concern will be discussed with the organization throughout the duration of the assessment, so that, near the end of the stage 2 audit, the organization should have a fair idea of what the final verdict will be.

A typical opening meeting would have on the agenda:

- Discussions of critical and non-critical findings from the stage 1 audit.
- Informing the organization that if problem areas are noted by the auditor, they will be briefly discussed at the time of the finding.
- Discussions of corrective action requests – what they mean and when they would be raised. The significance of and the differences between major or minor corrective actions are explained (see below).
- An explanation of the making of opportunities for improvement – such opportunities being items that may assist the organization to improve its EMS.
- The explanation that an external audit can only cover a sample of the EMS. Such a sample chosen by the audit team will be representative.

Audit progression

As previously stated, the objective is to verify system implementation by reference to compliance with the organization's own procedures. To achieve this, auditors will use pre-prepared checklists derived from the stage 1 desk-top study of the organization's documentation. Such checklists are used by auditors for making notes as the audit progresses, not only of evidence seen, but also for noting areas of non-compliance. Such non-compliances could lead to the generation of 'corrective action requests' (terminology may differ between certification bodies) and these are explained further below.

Corrective action requests

These are a fundamental part of the certification process as they are a mechanism to ensure continual improvement occurs over a period of time. Corrective action requests should be viewed as a positive outcome of the assessment rather than a negative one.

That said, the organization must demonstrate that it is following its own environmental policy and procedures. If areas of weakness are discovered then this will be noted. Such areas of weakness will result in a corrective action request being generated by the auditor. There are two categories of requests, depending upon the severity of the non-compliance.

- ***Minor corrective action requests***

  A minor corrective action is usually requested when a single observed lapse, an isolated incident, has been identified in the organization's EMS. The process is usually that the auditor will note the observed discrepancy on a dedicated form. This is handed to the organization's management representative. The management representative signs the document to acknowledge the observed lapse and this document is used to track progress of the corrective action. As an example, personnel may miss logging data in a register on one occasion. This is a case where a sound procedure was being followed the vast majority of the time but, perhaps for a reason originally unforeseen, it was not followed on this one occasion.

  Minor corrective action requests raised during an assessment do not necessarily prevent the organization being certified to ISO 14001. Once a minor CAR has been made, the organization is required to respond in writing (on the form mentioned above) within a specified period of time – typically 6 months. The organization should detail the actions taken or proposed, in order to prevent recurrence of the problem. The auditor can then respond to the organization to confirm, or otherwise, that the action or proposed action, is appropriate.

  This is a mechanism to ensure that the organization will take the appropriate corrective action, in advance of the next scheduled surveillance visit – for that is the occasion when the effectiveness of the corrective action can be fully verified by the auditor. It follows that sufficient evidence of the corrective action must have been generated to show the auditor on this visit. If, for whatever reason, there is insufficient evidence to demonstrate the effectiveness of the corrective action, the auditor will be placed in the position of escalating what was a minor corrective action request into a major corrective action request. This has implications for the continuation of the ISO 14001 registration.

- ***Major corrective action requests***
  Major corrective actions requested during a stage 2 audit will preclude the EMS from being certified. They act as a 'hold' point. A major corrective action is requested where there is an absence, or a total breakdown, of a procedure. It could be that at the stage 1 audit, that although a documented procedure was adequate to meet the requirements of the Standard, it may not in fact be fully implemented. For example, a small department in the organization may not have been given enough training in operating to a procedure. It may be that there is some personnel resistance to following the procedure. Whatever the reason, the system is not working as it was planned and documented. Additionally, where an observed non-compliance is, in the opinion of the auditors, likely to cause an environmental incident or accident, then the requested corrective action will always be categorized as major.

  Again, the auditor will note the discrepancy on a corrective action request form. However, on this occasion, a much shorter time-scale is allowed for a written response to be verified by the auditor on a dedicated extra visit. Such a visit is an additional expense to the organization. Typically, a written response is required within 1 month and the additional visit within 2 months from the date of the stage 2 audit. The purpose of this extra visit is purely to verify that the corrective action has taken place and that evidence to demonstrate this is available. If the major corrective action is satisfactorily resolved, then certification can proceed.

  The reason for this shorter time-scale is to ensure that the organization's momentum towards gaining certification does not falter. However, there should be sufficient time for the organization to strengthen its resources to correct the discrepancy.

There is much debate about what constitutes a minor or a major corrective action. There are guidelines within accreditation criteria and the certification bodies' own procedures, but in many respects the actual categorization depends upon the professional integrity of the team leader, or auditor, during the stage 2 audit who balances all the evidence available and then makes an objective judgement as to the category.

Opportunities for improvement

The auditors may also indicate that there are opportunities for improvement. These are intended to indicate to the organization areas of potential improvement and are generally followed through at the next surveillance visit. However, if the organization has considered such advice but decided against further actions – due to reasons that the auditor was not aware of – the auditor will not contest the point. Corrective action requests must be actioned – opportunities for improvement need not be.

The closing meeting

At the conclusion of the stage 2 audit, the auditing team will consider their individual findings and reach an agreement. The team leader will decide to do one of the following:

- Recommend certification unreservedly.
- Recommend certification with minor corrective actions.
- Delay certification until further work on the system is undertaken due to major corrective action requests.

Although a verdict is given to the organization by the lead auditor or team leader (as described previously) this is, strictly speaking, a recommendation only. The final decision to grant the award of the certificate must be given by the governing board of the certification body.

The report

The client receives a report usually issued at a later date describing in detail the progress of the assessment, the areas and departments, personnel and records that were audited and interviewed. Corrective action requests and opportunities for improvement are listed. This report is used by the implementing organization to action such deficiencies and by the certification body to trigger off the issue of the ISO 14001 certificate.

Surveillances, or continuous assessment audits, following the stage 2 audit

The assessment hurdle is not the end of the process – the organization must prepare for ongoing surveillances, or continuous assessment, depending upon the certification body's terminology.

Again, the approach described may differ slightly from one certification body to the next but some form of continual surveillance activity is required by the accreditation bodies. This surveillance activity must be, at a minimum, one surveillance visit every 12 months. These visits are planned by the certification body to ensure that over a period of 3 years the system is audited against every part of the Standard. At the end of such a 3-year period, the certification body reviews the overall continuing effectiveness of the organization's EMS. The review will cover the number and seriousness of corrective action requests, as well as any major changes to the organization's operations – including those which may require additional time on site at the next surveillance visit. (This is part of the certificate renewal process – see below.)

Whenever possible, the auditor for the surveillance visits will be a member of the team that performed the stage 2 audit. The auditor will therefore be familiar with the organization in question and will use the surveillance time more effectively. The objective of each surveillance visit is to sample only a part of the total system. However, at each visit the following is always examined:

- The effectiveness of the system with regard to achieving the objectives of the organization's environmental policy
- The level of management commitment
- The functioning of procedures for notifying authorities of any breaches of authorizations or discharge consents
- Progress of planned activities aimed at continuous improvement of environmental performance – where applicable
- The internal audit – methodology, scope, depth and follow-ups of any internally identified corrective actions
- Follow-up of any corrective actions raised by the certification body at the previous visit

Should there be no areas for concern at the surveillance visit, the auditor will generate a report, similar to the stage 2 audit report, for the organization. If applicable, the auditor will make recommendations to the organization for improvements to the system in the form of opportunities for improvement.

Certificate renewal

Depending upon the options chosen or offered by the certification body, some form of 'extra' scrutiny of the EMS is required every 3 years.

This can range from a prolonged surveillance visit to almost a full stage 2 audit visit. As described under *Surveillance methodology*, only a representative sample of the system is audited every 6 or 12 months by the auditor. It could be that, after 3 years, so many changes have occurred that the system, although appearing to be sound on the 'snapshot' type of review of a surveillance visit, may no longer meet all the requirements of the Standard.

Likewise, over a period of 3 years, many minor corrective action requests may have been raised indicating perhaps a lowering of the level of commitment by senior management to the principles of the Standard. In this case, a virtual full re-audit of the system is required to give the certification body a higher level of confidence. This option is usually included in the wording of the contract agreed between the certification body and the organization.

Certificate cancellation

An ISO 14001 certificate can be cancelled for several reasons. If an auditor finds evidence of deterioration of the EMS during a routine surveillance visit he may make a major corrective action request. Major corrective actions requested after certification are taken very seriously by certification bodies as they indicate a fundamental breakdown in the system – an EMS which the certification body has previously assessed and issued a certificate. The credibility of the certification body itself can be at risk.

If such a situation arises and the major corrective action request is not implemented within the required time-scale (usually 1 month, although there can be extensions to this) then the certification body will take the necessary steps to recall the certificate and cancel the registration to ISO 14001.

The certification body must be seen to be protecting the integrity of the certification process and the value of the certificate. The message must always be that ISO 14001 certification is not easily obtained and, once obtained, an organization must demonstrate sound environmental practice and continuous improvement in order to retain the certificate.

Summary

The certification process is highly regulated to ensure credibility and delivery of consistency to the implementing organization. The audit process has stages to ensure the correction of deficiencies step by step. Major areas of concern which could otherwise halt certification are thereby avoided.

By detailing the process, it is hoped that organizations about to implement ISO 14001 will know much more about how the certification bodies operate and will feel less daunted by the actual assessment process. The intention of any certification body must be to make the mechanics of the certification process easy whilst at the same time ensuring that where errors or deficiencies occur they are appropriately addressed through corrective action requests. Ongoing surveillance visits ensure that not only must certification be earned, it must be maintained.

Chapter 4

Integration of environmental management systems with other management systems

Introduction

In the first edition of this book other management systems were considered for the integration process such as the then BS7799 (Information Security Management) and Investors in People (IIP) and the chemical industry's 'Responsible Care Program'. However, although some organizations have successfully done this, in the majority of integrated systems certified to date, organizations have integrated their (usually) initial ISO 9001 system with ISO 14001. Moreover, in recent years there has been a tremendous surge of interest in Occupational Safety and Health management systems, and since the publication of OHSAS 18001 in 1999, this has stimulated further integration. Thus there are increasing numbers of organizations that now have a management system encompassing Quality, Environmental and Occupational Health and Safety management systems.

Thus in this second edition, only these three will be considered and this chapter discusses the relationship between ISO 9001, ISO 14001 and OHSAS 18001 as these standards represent the core business processes and form the basis of a 'total' management system for operating a successful business.

These three Standards were developed at different times and were designed to meet different needs and are aligned with each other in terms of:

- Intent to manage business risks
- Language and structure – in broad terms of clause numbering and sequence
- Philosophy – the concept of continuous improvement is inherent

This chapter will offer some guidance in management systems integration – an area undergoing rapid international development. Each organization must make its own choices as to whether to run a business with one or more separate systems or to join them together, i.e. integration.

Fundamentally, there is no reason why many separate management systems cannot be combined, amalgamated or integrated. However, with a book of this size, the emphasis is on ISO 14001 implementation and integration with other international standards. Likewise, this chapter will not list the clauses or requirements of each standard and compare and contrast them with ISO 14001. Instead, this chapter will focus on the common features of the systems. There are differences, however, and these will be highlighted as appropriate.

As a note, much development work is being done world-wide to develop such integrated management systems, and the ideal goal is to produce one definitive 'Standard' suitable for any business. This Standard would address all of an organization's activities and would be used as a model for the successful running of the business. The ISO Technical Management Board are investigating the possibility of producing an integrated management system 'Standard' which would effectively incorporate existing standards on occupational health and safety, quality assurance and environmental performance improvement.

Possible barriers to management system integration

Small- to medium-sized enterprise may hesitate to integrate their management systems. They may have stretched their resources to implement and achieve separate management systems in the past and may not now have the resources to 'knit' two or three management systems together. An increase of management time input in the short term could be difficult, even though there is no dispute over the longer-term efficiencies. So, integration may well be low on the list of their business priorities.

Organizations without any management systems may spend some time debating whether to implement systems separately or to take a bold step and seek initial integrated certification. Both routes have their advantages and disadvantages, and seeking advice on the best route may give conflicting opinions and delay any implementation decisions. Such organizations ask certification bodies questions about the integration process. Such questions can be categorized into the following set:

- Does having ISO 9001 certification assist in the process of implementation of ISO 14001?
- Is it just as easy to internally audit two systems as one?
- Are costs reduced because elements of the separate systems overlap?
- Will a non-conformance in one management system have a negative impact on the others?
- How well will the new system (ISO 14001 and/or OHSAS 18001) fit into, or combine with existing management systems – such as quality assurance, data protection and so forth?
- Will a whole new set of manuals and procedures be required, i.e. doubling or trebling the existing QMS system for example?

This chapter will address these issues and attempt to answer such questions.

Definition of an integrated management system

Within the scope of this book, an integrated management system is a management system comprising of ISO 14001 plus at least one other management system. Both (or more) systems should run concurrently with each other in an organization and both should be capable of being audited by an external body to a recognized national or international standard.

Different ways of managing this integration are discussed and several 'models' put forward. None of these models are right or wrong. The extent, depth and breadth of integration are decided by each individual organization.

Note that an integrated management system does not take the place of, or compete with a Total Quality Management (TQM) system. TQM seeks changes in the culture of the organization and fundamental reappraisal of business practices and can be quite revolutionary in its outcome. Integration is still concerned with elements of 'compliance to procedures' and is somewhat less revolutionary in philosophy, as it considers a narrower set of criteria of running a business such as:

- Management responsibilities and accountabilities
- Business processes
- Deployment of resources, skills, knowledge and technology

These aspects are integrated to ensure that the business delivers its objectives. The objectives of the business include elements of quality, the environment and occupational health and safety. These objectives are the same as many of the requirements of stakeholders:

Customers require:	**Shareholders require:**
Safe and reliable products/services	Return on investment
Reliability of supply	Profitable business
Fitness for purpose	Legal compliance
Environmentally safe product	Good image
Value for money	Growth

Employees require:	Community requires:
Safe working environment	Minimum environmental impact
Job satisfaction and security	Employment opportunities
Care and recognition	Stability
Rewards for good work	

In many ways such requirements are not met by just environmental responsibility but a much wider range of employer care and responsibility.

Reasons for integration of management systems

Organizations implementing integration respond with two predominant answers when asked about their reasons for seeking an integrated management system:

1) To reduce costs to the business and add value to processes

The costs referenced here are related to the most efficient use of management time. This includes better use of auditors' time (both internal auditors and certification bodies auditors). The reduction in management time has tremendous internal cost benefits such as more efficient maintenance of the management systems. The burden on management time within the organization can be reduced if one element of a management system can be addressed at the same time as the same element of the other system. For example, the environmental management review for ISO 14001 could take place at the same time, with the same personnel, as the management review for ISO 9001.

Fees incurred for the Certification Body to carry out its routine surveillance/ continuous registration visits can be minimized by only having the external auditor arriving every 6 months – under some accreditation systems – (that is, twice a year to perform a combined ISO 9001 and ISO 14001 audit rather than four visits, two for ISO 9001 and two for ISO 14001). As suggested in the introduction, a single standard that integrates all elements of a modern management system into one auditable standard is the ideal. However, for the near future, it must be appreciated that individual standards have been developed for different purposes and therefore organizations wishing to integrate their management systems will need to develop their own model for an integrated system.

'Adding value' was a reason also given, and a definition can be to ensure that the activities and processes of operating a management system, have a positive and measurable impact upon the profit and loss accounts of a business. Note that:

- 'Added value' within management systems has a history. As ISO 9000:1994 matured, concerns were raised world-wide that this quality standard should be the driver for improvements to products and services, rather than just a control tool.

- Certification bodies were questioned by industry and commerce as to their ability to add value to ongoing audits. This added value meaning passing on positive ideas for improvement rather than just auditing for compliance to the Standard and documented procedures, leaving negative comments behind in the form of corrective action requests.

- Certification Body auditors were certainly well placed to offer such ideas for improvement, as throughout their auditing careers they had been in a unique and privileged position of having access to a wide variety of organizations, cultures, processes and business sectors reviewing both inadequate practices, but more significantly having exposure to 'best practices'.

- It was generally felt that if clients of certification bodies were not making more profits or becoming more secure commercially, than they would be without certification, then no value was being added. Added value was often seen as getting closer to understanding what a company's key business issues were. Since those days of course, ISO 9001:2000 demands an approach of continuous improvement with setting of quality objectives and/or business objectives, and although certification bodies will still offer opportunities for improvement based on best practices, in a sense this original definition and debate of added value has become redundant. The concept and understanding of added value has undergone many changes, and in the context of this chapter refers to the value obtained by integrating two or three management systems, i.e. the synergy obtained by the resultant integrated management system.

2) To reduce risks to the viability of the business

The management of an organization may well perform an analysis of the risks to the business. Many forward thinking organizations look at three components and question:

- *Quality*: what are the risks of supplying a product or service that does not meet customers' requirements and more to the point does not keep up to date with these changing and more demanding requirements (the concept of continuous improvement).

  ISO 9001 is the tool for reducing this risk.

- *Environment*: what are the risks of not complying with legislation? If the organization does not keep up to date with best practices of environmental management? What risks are being taken in terms of adverse publicity towards 'household names' etc. if the organization is perceived to be of poor environmental probity?

  ISO 14001 is the tool for the reduction of these risks.

- *Occupational health and safety*: what are the risks of causing injury to the workforce through the use of out of date or negligent practices? These risks include at the least lost time at work – therefore productivity suffers – to civil or criminal proceedings from injured personnel, costing time and financial penalties as well as adverse publicity.

  OHSAS 18001 is the tool for managing these risks.

Such organizations know that there are always new challenges and demands to be met when managing any business, especially when viewed against:

- Significant competition
- High customer and community expectations

- Returns on capital employed
- Regulatory compliance
- Executive liability risk

If an organization is to meet the challenges above successfully, then it needs to call on all its management resources – especially when fulfilling the requirements of Quality, Health and Safety and Environmental Standards. Integration of these management standards would allow common areas of the standards to be managed, thereby making more effective use of management time and addressing the above challenges. Looking at the implications and consequences of a separate systems approach:

- Actions and decisions are made in isolation and are therefore not optimal.
- Employees are presented with a proliferation of information and even conflicting instructions which may put the company at risk.
- Bureaucracy can flourish – how many systems can the organization cope with?
- Lack of ownership.
- Wrong and costly decisions are made due to non-optimization of resources.

These consequences can be addressed by developing an integrated approach to quality, health and safety, and environmental management in particular as these three standards cover a high proportion of the scope of any business. Integration will expose areas of waste and no-value-added activity, and provide opportunity for rationalization of:

- Documentation
- Auditing and review processes
- Barriers across departments and functions (by their removal)

Integration – brief overview of ISO 9001:2000 and OHSAS 18001

These two Standards are the most likely 'candidates' to be integrated with ISO 14001.

ISO 9001:2000 is the Standard for the management of quality. The majority of organizations implementing ISO 14001, or who have gained certification, generally have been operating a quality management system for some years. Due to its long history of development (some 30 years or so) there are many excellent books available covering all aspects of quality assurance. However, some description would be useful at this point for comparison purposes with ISO 14001:

Briefly, in order to be successful an organization must offer products or services that:

- Meet a well-defined need, use or purpose
- Satisfy customer expectations
- Comply with applicable standards and specifications
- Are made available at competitive prices
- Are provided at a cost which will yield a profit

ISO 9001 acts as a model for an organization to follow in order to meet these requirements.

OHSAS 18001:1999 is the Standard for the management of occupational health and safety. It is based upon hazard identification and risk assessment, i.e. the management of risk, which sets it apart from previous national and international safety standards.

In recent years organizations of all types have been actively addressing the question of health and safety in the workplace, partly in response to legal requirements but also in response to the fear of litigation and the pressures of increasing insurance premiums.

Adverse publicity associated with unsafe practices, realization that accidents have negative cost implications for organizations' money through stoppages, investigations, loss of productivity are all drivers for better safety controls.

True assurance on safety can only be achieved in the context of a structured management system, which strives to manage safety in an active way rather than simply to react to events.

As yet, it is not a true 'ISO' Standard, but has gained international acceptance very quickly.

Table of comparisons of the three standards

Prior to integration of one of more of the Standards, it would be useful to review the following table and also Figure 4.1 which demonstrate the commonalities in structure, approach and philosophy of the Standards.

Clause ISO 14001		**Clause ISO 9001**		**Clause OHSAS 18001**	
1	scope	1	scope	1	scope
2	normative references	2	normative references	2	reference publications
3	terms and definitions	3	terms and definitions	3	terms and definitions
4	EMS requirements	4	quality management system	4	OH&S management system elements
4.1	general requirements	4.1	general requirements	4.1	general requirements
4.2	environmental policy	5.1	management commitment	4.2	OH&S policy
4.3	planning	5.4	planning	4.3	planning
4.3.1	environmental aspects	5.2	customer focus	4.3.1	planning for hazard identification, risk assessment and risk control

Clause ISO 14001		Clause ISO 9001		Clause OHSAS 18001	
4.3.2	legal and other environmental requirements	5.2 7.2.1	customer focus requirements related to the product	4.3.2	legal and other requirements
4.3.3	objectives, targets and programme(s)	5.4.1	quality objectives	4.3.3	objectives
4.4	implementation and operation	7.0 7.1	product realization planning of product realization	4.4	implementation and operation
4.4.1	resources, roles, responsibilities and authority	5.0 6.0	management responsibility resource management	4.4.1	structure and responsibility
4.4.2	competence, training and awareness	6.2.2	competence, awareness and training	4.4.2	training, awareness and competence
4.4.3	communication	5.5.3 7.2.3	internal communications customer communication	4.4.3	consultation and communication
4.4.4	documentation	4.2 4.2.1 4.2.2	document requirements general quality manual	4.4.4	documentation
4.4.5	control of documents	4.2.3	control of documents	4.4.5	document and data control
4.4.6	operational control	4.7	product realization	4.4.6	operational control

Clause ISO 14001		Clause ISO 9001		Clause OHSAS 18001	
4.4.7	emergency preparedness and response	8.3	control of non-conforming product	4.4.7	emergency preparedness and response
4.5	checking and corrective actions	8.0	measurement, analysis and improvement	4.5	checking and corrective action
4.5.1	monitoring and measurement	7.6	control of monitoring and measuring devices	4.5.1	performance monitoring and measurement
		8.1	general		
		8.2	monitoring and measurement		
		8.2.1	customer satisfaction		
		8.2.3	monitoring and measurement of processes		
		8.2.4	monitoring and measurement of product		
		8.4	analysis of data		
4.5.2	evaluation of compliance	7.2.1	determination of requirements related to the product	4.5.1	performance monitoring and measurement
4.5.3	non-conformity, corrective and preventive action	8.3	control of non-conforming product	4.5.2	accidents, incidents, non-conformances and corrective and preventive actions
		8.5.2	corrective action		
		8.5.3	preventive action		
4.5.4	records	4.2.4	control of records	4.5.3	records and records management
4.5.5	internal audit	8.2.2	internal audit	4.5.4	audit
4.6	management review	5.6	management review	4.6	management review

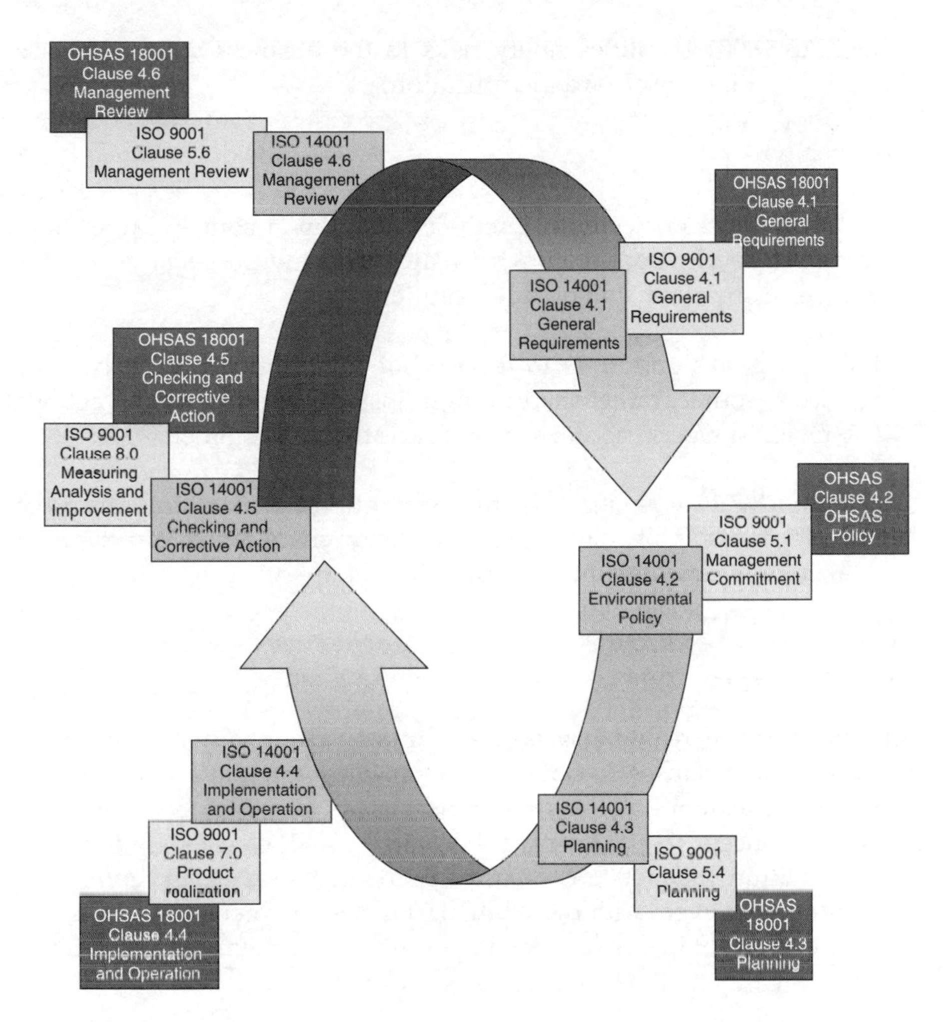

Figure 4.1 Integrated implementation cycle for continuous improvement

Some commonalities

Some commonalities are readily apparent from Figure 4.1 and the above table. Some clauses are numbered the same, some are titled the same. The underlying philosophy is the same. For example:

- ISO 14001 identifies environmental risks to the business through clause 4.3.1 (environmental aspects)
- ISO 9001 identifies commercial risks to the business through clause 5.2 (customer focus)

- OHSAS 18001 identifies safety risks to the business through clause 4.3.1 (planning for hazard identification)

Additionally:

- ISO 14001 seeks objectives to reduce pollution at source (rather than treat or recycle pollution, which are in many respects corrective actions or so-called 'end of pipe' solutions).

- ISO 9001 seeks objectives to reduce out of specification products or service at source, rather than through inspection or corrective actions to prevent such non-conformances reaching the customer.

- OHSAS 18001 seeks objectives to design out hazards in the workplace rather than using personnel protective equipment for example, which again can be considered to be corrective actions.

Some differences

Differences are in reality few and are apparent in the above list or in Figure 4.1. Some clauses do not have exact alignment to each other either in number, or text, or both. At first glance, it may be concluded that there are some conceptual differences between some clauses. Emergency response within ISO 14001 can easily be related to a major spillage of chemicals into a river. Within OHSAS 18001 the emergency response can again be thought of in terms of dealing with a major chemical spillage, but this time threatening the safety of the workforce.

ISO 9001 does not appear to have a similar scenario. However, if it is considered that the ISO 9001 certified organizations manufacturing vehicles, medical devices, drugs, toys and so on, then if a critical, or life-threatening manufacturing 'fault' is discovered, the product may have to be withdrawn from the market very rapidly. This would certainly be an emergency and plans put in place, and tested, where appropriate.

Considering these standards, therefore, there are more commonalities than differences. As described below, there is a spectrum of 'degrees' of integration, and it is recommended that an organization plan and decide at what level integration will work for them. The models suggested below are simplified and other models may well exist.

Models of integration

Model 1: At 0% integration, there could be all three systems, such as documented policies and procedures very clearly defined, for example, held in one hard-copy folder, or within an electronic system, in a named directory. It is very obvious where one system starts and ends with no overlapping of procedures or management processes. It would be very unlikely that changes in one system would impact upon the others. The level of integration may go as far as keeping the documentation, hard copy or electronic, in the same location under the control of one manager.

Three management systems may well exist, under the control of three separate management representatives with little or no overlap in terms of meetings, reviews etc. Some documents at level 4 may be common such as training request forms or calibration record forms (see Figure 4.2).

Model 2: At 50% integration, the appropriate policies and procedures are now held in one folder or electronic directory, but now each process or procedure has three recognizable elements to it, – ISO 9001 considerations, ISO 14001 considerations and OHSAS 18001 considerations. There will probably still be three sets of objectives and targets. The level 4 forms and records will probably be the same and the level 3 documents, such as work instructions, may have all three elements of the Standards within one

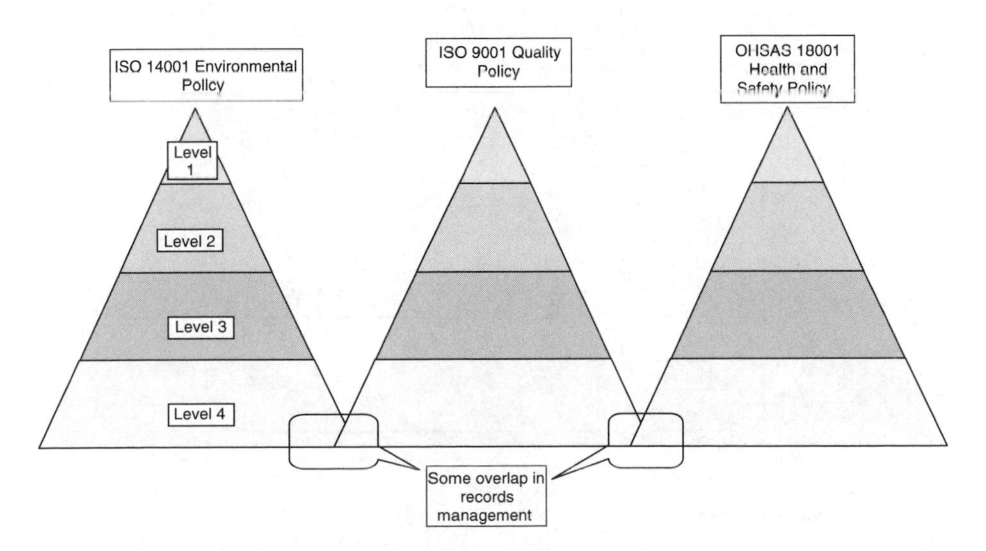

Figure 4.2 Model 1 integrated documentation structure

instruction. Some objectives and targets will overlap. One manager may well co-ordinate all three systems but it is still fairly transparent where one system stops and another one begins (see Figure 4.3).

Model 3: At 100% integration, the boundary between the Standards are seamless and procedures describe processes in terms of risk management and, for example, the management system could only be dissected into its three components with difficulty. There is now one overall policy for the site and all the objectives and targets are contained within one set of common objectives (see Figure 4.4).

As an illustrative example, Figures 4.5, 4.6 and 4.7 show a spillage process clearly identified to meet the requirements of ISO 14001, ISO 9001 and OHSAS 18001, respectively. Figure 4.8 shows an integrated spillage process addressing the requirements of all three standards.

Views and opinions of integration

An organization may decide that the integration of its separate certified management systems is necessary from both a use of resources and costs

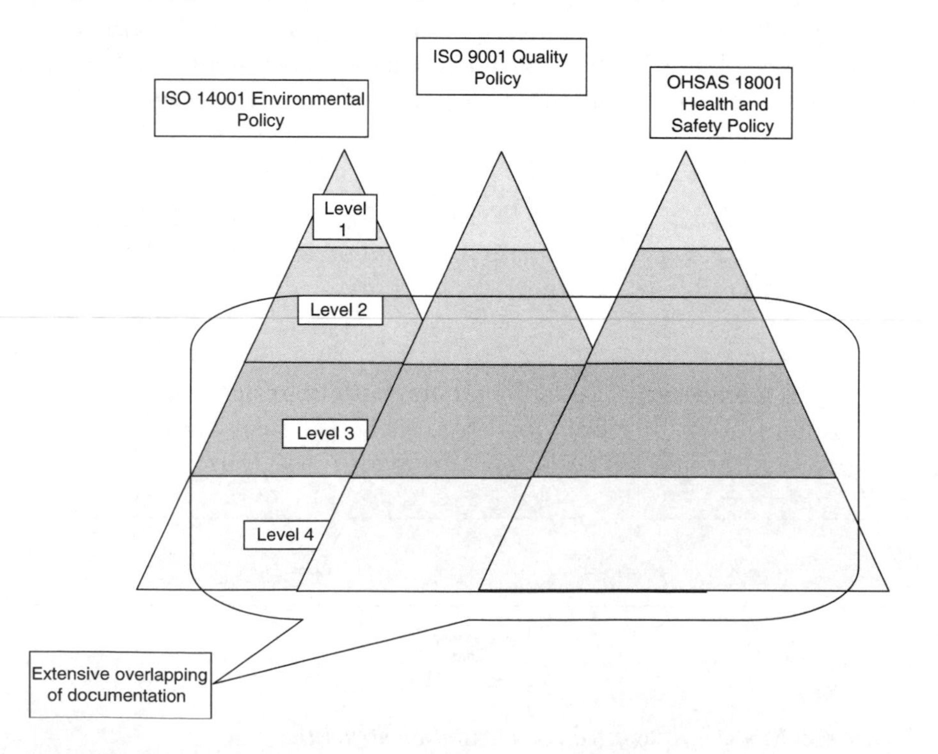

Figure 4.3 Model 2 integrated documentation structure

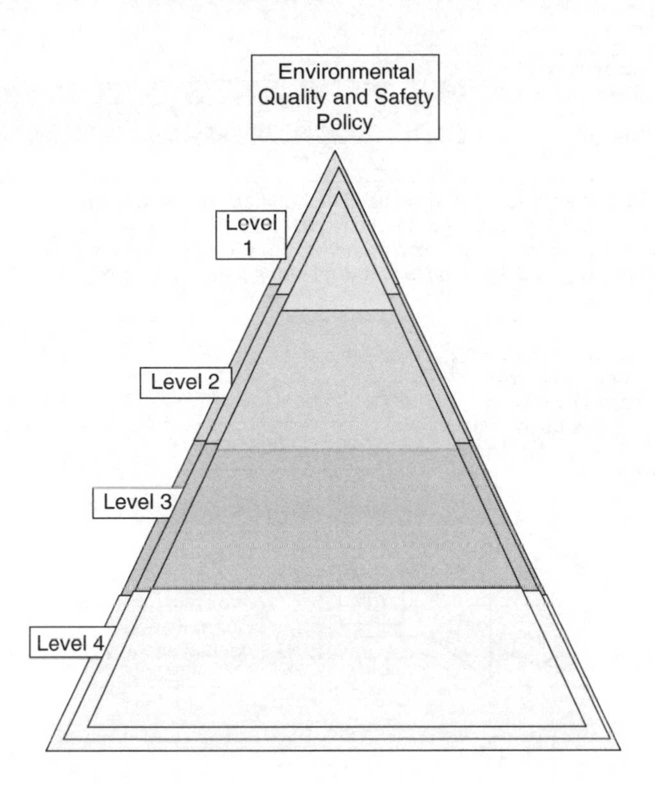

Figure 4.4 Model 3 integrated documentation structure

point of view. It makes sense therefore to understand the stance, expectations or opinions of the various interested parties, i.e. stakeholders.

ISO 14001 – the Standard

The introduction in the Standard appears to encourage integration by referring to 'integration with other management requirements to assist organizations achieve environmental and economic goals'.

The certification bodies

Integration is encouraged by certification bodies, as it reduces certification costs (due to combined audits) and less duplication of documentation for the client.

However, the practicability of simultaneous certification is dependent upon the correct mix of auditors being available. There are logistical problems getting the right mix of auditor skills together at the right time and place. The development of the multi-discipline auditor (see Chapter 6) may ease this practical limitation in the longer term.

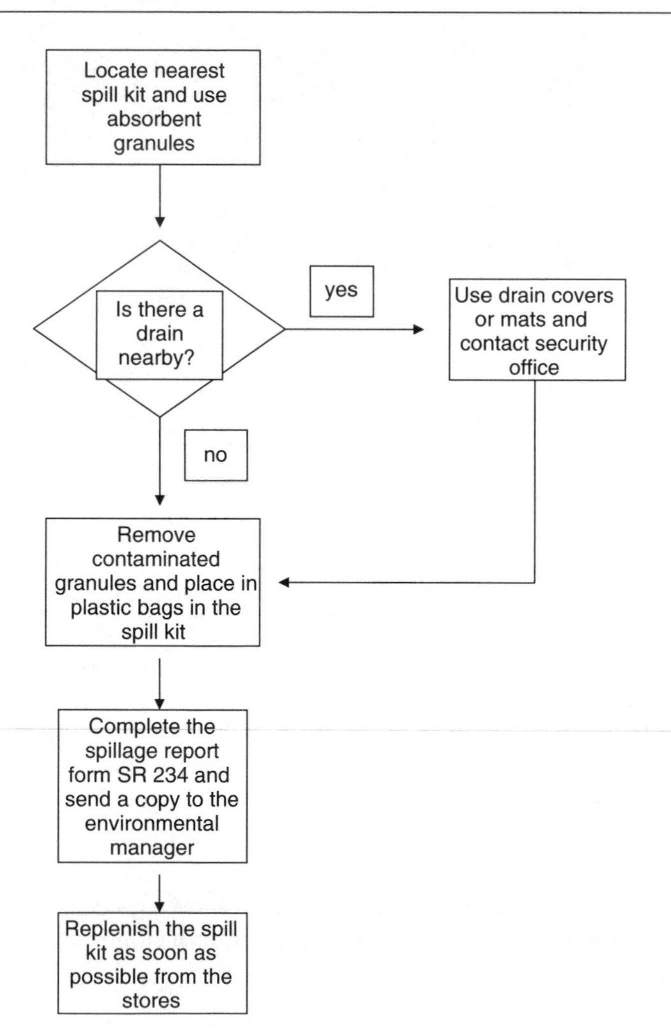

Figure 4.5 A spillage process meeting the requirements of ISO 14001

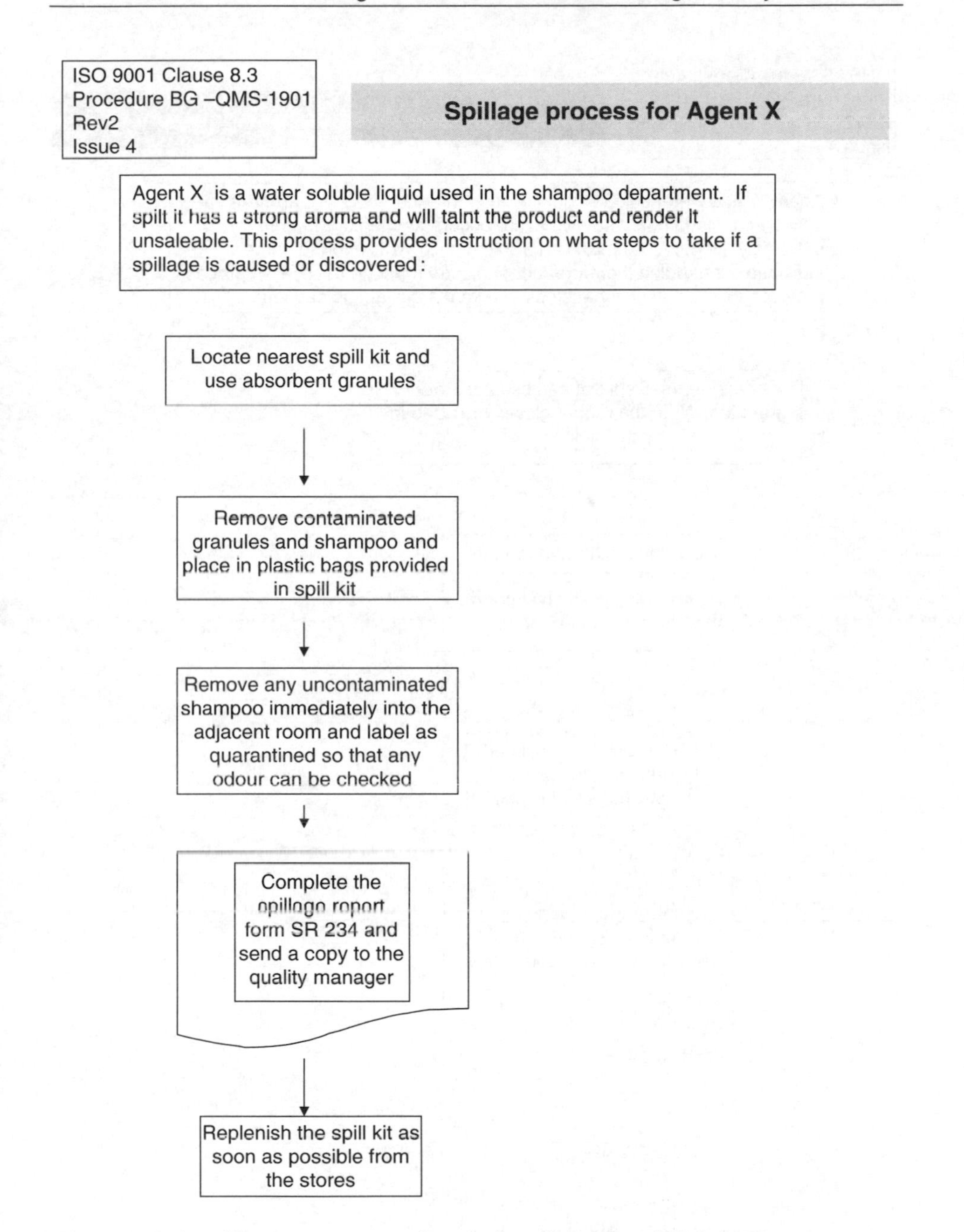

Figure 4.6 A spillage process meeting the requirements of ISO 9001

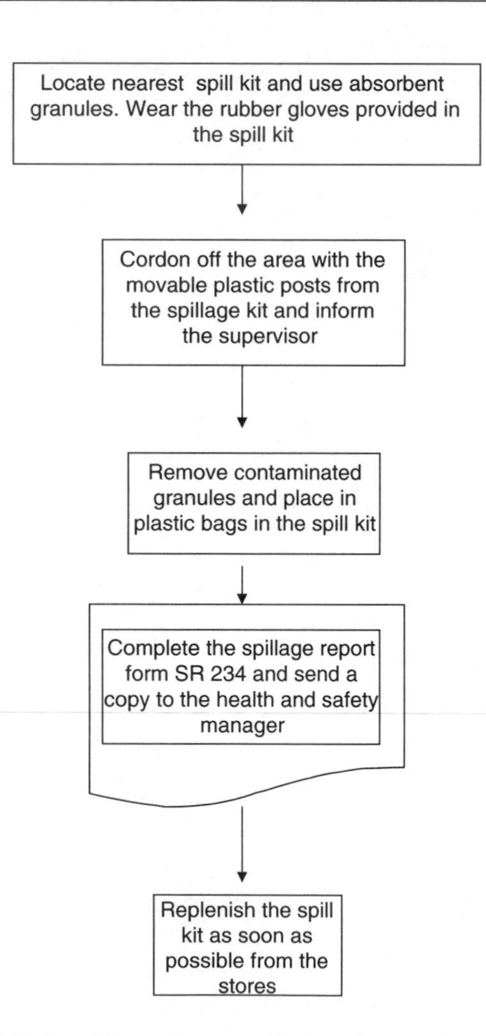

Figure 4.7 A spillage process meeting the requirements of OHSAS 18001

ISO 14001 & OHSAS 18001Clause 4.4.7/ ISO 9001 Clause 8.3
Procedure BG –IMS- 1904
Rev2
Issue 4

Spillage process for Agent X

Agent X is a water soluble liquid used in the shampoo department. If spilt and enters the surface water drains, it has a very high COD, may damage aquatic life and will be a breach of the site discharge consent.
Additionally it has a strong aroma and will taint the product and make it unsaleable. If spilt onto the tiled floor will cause a safety hazard as it is very slippy. Some individuals may become sensitized so personal protective equipment such as rubber gloves must be used. This process provides instruction on what steps to take if a spillage is caused or discovered:

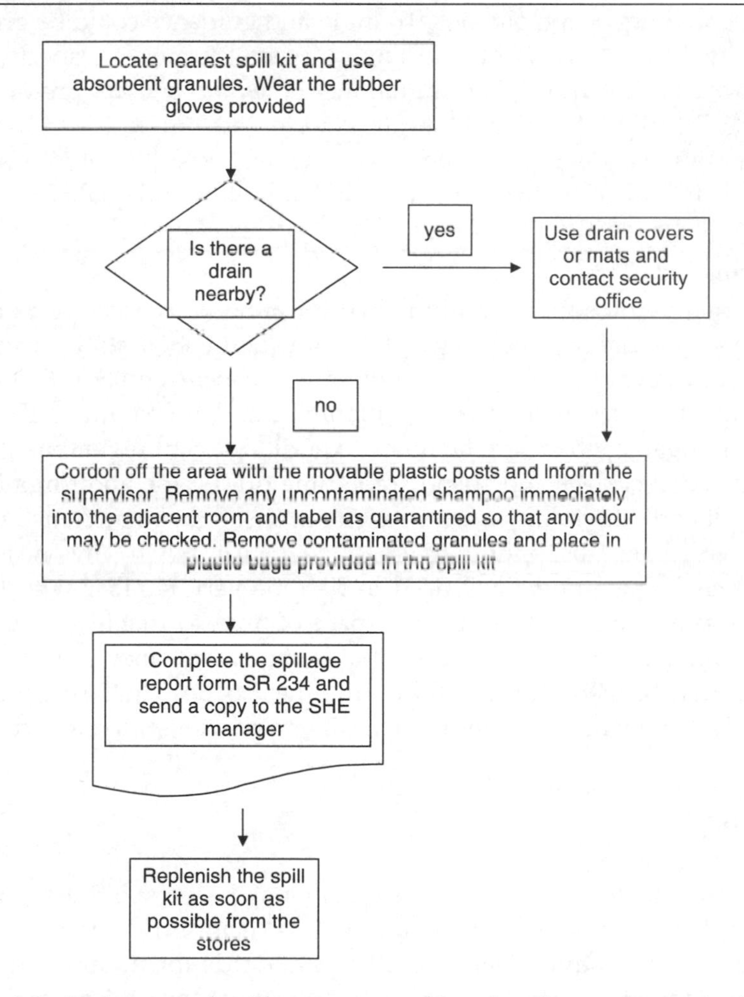

Figure 4.8 An integrated spillage process addressing the requirements of all three standards

ria (see Appendix II) which gives guidance for
the suggestion that the stage 1 and stage 2 audits
ıdits of other management systems. The criteria
the EMS should be readily identifiable in certifi-
d reports so that the quality, or the integrity of
ely affected by the combination of the audits.

...uditor

For the internal auditor the requirement to audit an integrated management system should not be too onerous. Training and guidance could be given by the quality, health and safety and environmental 'experts' respectively in house to the auditor via briefing or guidance notes and perhaps pre-prepared checklists. Usually, internal auditors are experienced individuals from within the organization and a programme of training may be sufficient to raise their level of awareness of environmental, health and safety and quality issues.

The external auditor

The external auditor representing the certification body may find an integrated system more of a challenge. At the time of the ISO 14001 certification audit, it may be that an organization requires a concurrent ISO 9000 assessment. The challenge here is one of planning and preparation. The team leader can organize the team members to audit different requirements with pre-prepared checklists and spend some time during the audit monitoring progress of both 'disciplines'. A greater challenge might be presented during routine surveillance, or continuous assessment, visits when only one auditor might attend. The auditor will need to become very familiar with two or more management systems in a short space of time, to enable a meaningful audit to take place and also to add value to the organization's system. This challenge may be made harder if the auditor has not been to the organization previously – either at the certification audit or a previous surveillance visit.

Summary

This chapter has examined the current management standards that are being implemented in a number of business sectors. The standards covered are not exhaustive. There are other standards for training, information management, forestry management and social and ethical sourcing of materials. However, the standards that are the focus of this chapter are the most widely used by the business community.

Chapter 5

Case studies

This chapter consists of contributions from the following organizations.

Case Study 1) SSL International plc

This organization's case study was included in the first edition of this handbook in 1997. The case study is included in this second edition as it illustrates the processes of continuous improvement and how such momentum can be maintained over a longer period of time.

Case Study 2) Lex Vehicle Leasing Ltd

This organization is an example of a service organization providing the consumer with an environmentally emotive product – the car – and the unique challenges overcome during the 'road' to ISO 14001 certification.

Case Study 3) Waste Recycling Group Ltd

This case study is included as an example of a service company whose core business is the removal, segregation, storage, destruction, incineration and burial etc. of the waste produced by our consumer society. Such activities of course produce their own set of environmental impacts and for which there is a whole raft of legislation and strict enforcement regimes controlling all aspects of this business.

The organizations above have successfully implemented ISO 14001. They are very different in terms of the sector they operate in and in the stakeholders to whom they are responsible. The content of each of the case studies varies which is not only a reflection of the different contributors' styles but also the fact that each of these organizations faced their own particular challenges during implementation. Hopefully, the reader will be able to identify with some of these challenges and adapt the solution to their own set of environmental circumstances.

SSL International plc

Written by Ingrid Hogan, Group Quality and Environmental Manager, SSL International plc, Tubiton House, Oldham, United Kingdom

Brief historical note

SSL International plc (then Seton Healthcare Group) registered its first manufacturing site to BS 7750, now ISO 14001, back in 1996. In those early years the company had approximately 1000 employees, three manufacturing sites, one distribution centre and its operations were mainly based in the north west of the UK.

The product range consisted primarily of textile-based products such as tubular bandages and athletic supports; wound care products such as irrigation solutions, swabs/wipes and dressings and a mix of over-the-counter products in the form of tablets, liquids, creams and ointments.

Since that time the company has grown significantly in size, geographical spread of operations and product range. It now has over 7000 employees, 27 manufacturing and distribution sites (including two joint ventures) located not only in the UK but also in continental Europe, Asia/Pacific and the Americas.

The product range has been extended to include condoms, medical gloves, household/industrial gloves and footcare/footwear products.

Ongoing change

This high level of ongoing change has meant that the environmental management system has had to be a flexible one, adapting and changing as the company evolved. With each acquisition/merger new sites have been added to the ISO 14001 implementation programme. The first two sites to become registered to ISO 14001 were the Oldham (UK) manufacturing site and the distribution warehouse at Middleton, not far from Oldham. In those early days there was very little in the way of in-house expertise, off-the-shelf training courses or best practice examples, so consultancy support had to be used throughout the implementation process. One of the first actions to be undertaken was to identify a course that would provide the Quality and Environment Manager with a minimum understanding of environmental issues. There were very few around and the only suitable one was a basic course in environmental auditing. Implementation was very much a learning process for all involved.

The implementation process began by looking at site activities, department by department. For each department relevant legislation was identified and a list of significant impacts drawn up. These lists were then reviewed and a set of environmental objectives agreed. At the same time core procedures were written to cover management system elements such as Environmental Aspects Evaluation, Environmental Auditing and Management Review.

All this took 12 months to complete. However a pre-assessment by the chosen certification body, revealed that the departmental approach had produced a fragmented system that did not really address the main issues and did not demonstrate a clear link between significant impacts, operational control and continual improvement. Work then began to address these observations. The methodology for evaluation of site activities was changed from a departmentally focused approach to one related to processes. One coherent list of significant environmental impacts was produced for the site, enabling cross-departmental procedures to be written for operational control and one set of site environmental objectives to be set.

Once the Oldham and Middleton sites had obtained their registration, and in line with the company's stated objective to extend the management system to all sites, implementation was rolled-out to other UK sites, one by one.

Changes in system philosophy

During the roll-out process it became obvious that environmental management was a shared responsibility. When it came to implementation at the fourth UK site (Bootle) a new approach was adopted, one that moved away from a designated individual being the site environmental expert to one actively involving all site management.

The preliminary review of site activities was still completed by the Quality and Environment Manager but, for the first time, this was followed-up by a training session for all site management, involving them in the decision-making process and assigning them responsibilities. The half-day training session began with a presentation of some background information on environmental issues, an overview of ISO 14001 requirements and details of the company approach towards ISO 14001 implementation. This was followed by a discussion of the results of the preparatory review, focusing on the site's significant impacts and the action plan proposed by the site Management Representative. Managers were then asked how they felt they could contribute to the implementation process, and the action plan was revised accordingly and responsibilities assigned. This effectively ensured that site management became the project team. Following on from the initial training session regular meetings were held to discuss progress and agree further actions.

This approach proved very successful in ensuring the commitment of everyone on site. It changed the culture and ensured that environmental considerations became fully integrated with everyone's routine activities. However, because it still relied on Group personnel to complete the preliminary review of site activities and help define an initial action plan, it was not suitable for SSL International sites outside of the UK.

The challenge for the Company's sites outside of the UK

In 1999 implementation began at the two SSL Malaysian sites. This posed a challenge, not least because Group personnel had no previous knowledge of Malaysian legislation or the sites' manufacturing processes. In addition, the two sites were much larger than any UK site, with approximately 1000 employees each. Building up the expertise of local personnel was therefore critical and training became a key issue.

Three types of training were identified: ISO 14001 and impact evaluation for the project team; more detailed training covering environmental issues, legislation, management and auditing for the site Management Representatives; and general awareness training for the workforce. It was decided that Group personnel would provide training for the project team, external training would be sought for the Management Representative and site personnel would deliver awareness training later on in the implementation process.

In preparation for training the project team, a scoring system was developed for the environmental impact evaluation procedure to make it more user friendly and less open to subjective interpretation. A 2-day workshop was held at each site to show personnel how to use the updated procedure to review site activities, identify relevant legislation and significant environmental impacts, complete a gap analysis and draw up an action plan.

Challenges of training

Training SSL International plc Malaysian personnel was very different to providing training for UK site management. In the UK, site personnel could be relied on to ask questions when they were unsure of anything. The factory 'culture' in Malaysia was different in that staff did not readily question management decisions or initiatives. Thus a more interactive style of training had to be developed. Each site had to take one process and complete a full evaluation as part of their training. At the end of the 2 days they then had to give a presentation on their findings and this was used to assess how well they had understood the procedure.

When identifying an external course for the site Management Representatives, it was important that a course was found that ensured training was equivalent to that found in the UK. Fortunately this was not difficult as the Institute of Environmental Management and Assessment (IEMA) certified training courses world-wide. These courses had been used to train UK personnel and were available in Malaysia. This provided the site Management Representatives with a good working knowledge of Malaysian legislation and environmental best practice: essential, as Group personnel could not provide this. Finally, towards the end of the implementation process, induction and awareness training packages were developed and delivered by site personnel, who had by this time gained much expertise in what were the pertinent issues for the site.

Ongoing maintenance and continual improvement

SSL International plc's ISO 14001 environmental management system is now fairly extensive with 12 sites already registered to the Standard and a further four currently in the implementation phase. There are several others in the very early stages of aspects evaluation. For Group personnel, helping sites put in EMSs has become routine and the focus is now on actively driving continual improvement throughout all areas of the company. Change has meant that it has sometimes been difficult to maintain progress with improvement projects as priorities for the company have shifted with each new acquisition or merger. This does not mean, however, that there have been no success stories and when a matrix is drawn of the company's achievements since 1995/6 it is evident that there has been a demonstrable improvement in environmental performance overall, although targets have changed with time.

Examples of success and failure due to changes in the business

One example of an early objective that was affected by changes within the company was the objective to increase the use of recycled packaging. In 1995 a target was set to replace all virgin board used for hand-packed cartons with recycled board. In the first year of the project the percentage had risen from 29% to 41.5%. Thus by the second year of the project, the target had been achieved. A success story!

However, when the original target for percentage of recycling had been defined, a large percentage of all products were hand packed into cartons and the target was appropriate and achievable.

As the Company grew in size, due to other acquisitions, more and more product ranges acquired were packed on automated packing lines. Such lines could not run effectively using recycled board so therefore they began reusing virgin board. This meant a new target had to be set to replace all virgin boards used for machine-packed products and the search was on to find a suitable alternative. (By 1998/9 only 54% of the product range was hand packed and despite running machine trials with several types of recycled board no suitable alternative could be found.)

Changes due to rethinking best environmental options

By 1999 increased knowledge of the respective life cycles of virgin and recycled boards cast doubt on whether recycled board was necessarily the best environmental option. An investigation comparing the two boards was commissioned. The conclusion was that there was little to choose between the two and that, because virgin fibres were stronger than recycled fibres, it would be possible to make cartons thinner and reduce overall consumption of resources by changing back to virgin boards. Company policy was therefore reversed and now all cartons purchased are made from virgin board, sourced from sustainable forestry operations to minimize environmental impact.

A further example of an objective that changed focus after early success was the objective to improve fuel consumption in car and commercial fleets. Initially, targets were set to achieve a 10% improvement by 1997/8 and a further 5% by 2000/1.

Company cars were numbered in their hundreds and so appeared as a significant environmental impact. The first step was to review historical data to determine the best approach to take. Changes were made to the car fleet and a 10.3% improvement achieved was 12 months ahead of schedule. A further review of company cars identified one type and model of car as the best environmental option and after making this the Company car of choice for the Sales Team, a 15% target was achieved 2 years ahead of schedule.

However, by 2000 the company had undergone two mergers and fuel use in company cars accounted for a much smaller percentage of business travel than it had in the past. The project was refocused to address the issue of business travel as a whole, particularly relevant as the newly merged company was more international in its operations.

Money was invested in the provision of video conference facilities, however, it proved very difficult to quantify fuel savings from the use of these facilities and attention was diverted into finding a way of monitoring business travel as a whole. Much management time was taken up trying to identify individuals world-wide who were responsible for booking travel and to finding out what travel records were being kept.

In the end it was felt that management time would be better spent and more environmental benefit would be gained from refocusing on the use of fuels consumed in manufacturing processes. In fact one site in Malaysia achieved a 35% reduction in fuel oil use and overall fuel oil use was down from 220 million kWh in 1999/2000 to 90 000 million kWh.

One final example is the environmental objective to remove ozone-depleting substances in the supply chain. This time it was target dates that were changed as problems were encountered during the course of the project. The initial targets were to replace CFCs used in aerosol products and HCFCs used as blowing agents in the manufacture of polyurethane foam used in wound dressings. The first target was achieved on time in 1995/6 with no serious problems encountered. The same could not be said with the polyurethane foam project.

When the project was first established a completion date of 1997/8 was set. Initial trials in 1996/7 with a foam using methyl chloride as a blowing agent were very successful, with sample dressings passing preliminary tests. A whole series of samples were then produced for further evaluation. All passed, except the last set that were tested for market acceptability.

User comfort versus environmental improvements
These were judged by users of the product to be too rough and abrasive against the skin and the project had to be put back to the design process to source a softer foam – delaying the target date by about 12 months.

Setbacks
Halfway through the process of finding an alternative softer foam, one of the foam manufacturer's key suppliers had to be replaced. This further delayed the project as a new supplier had to be found and additional testing completed on the raw material in question to ensure that it was in no way different from that which the previous supplier had provided. The project was delayed for a further year but eventually a set of sample dressings was produced that passed all tests. The objective was finally achieved, 2 years later than the original date.

The above experiences demonstrate that, although improvements have occurred, they have not always been achieved in the planned timescales. Success in environmental progress is therefore best viewed over a good few years of implementation and operation to see true continuous improvement.

Conclusion

SSL International plc has changed significantly from the time it was first registered to ISO 14001 in 1996 and its EMS has adapted accordingly. Many of the original environmental objectives still remain but they have changed focus over time and new ones have been added as new impacts have been identified.

SSL International plc has a mature EMS and increasingly it is looking outwards to identify opportunities to minimize its environmental impact further along the supply chain. The most recent objective is to promote local initiatives with stakeholders and the development of a methodical approach towards stakeholder dialogue.

Environmental achievements 1995–2002

Objective	1995/6	1996/7	1997/8	1998/9	1999/2000	2000/1	2001/2
To extend EMS to all manufacturing/distribution sites	Oldham and Middleton sites (UK) registered to BS 7750	Registration converted to ISO 14001, implementation begun at Redruth (UK)	Redruth site registered, implementation begun at Bootle (UK)	Implementation ongoing at Bootle and begun at Derby (UK)	Bootle site registered, implementation ongoing at Derby and begun at Rubi (Spain) and Kulim (Malaysia)	Rubi site registered, implementation ongoing at Derby and Kulim and begun in Portugal, Thailand, India and Heywood (UK)	3 sites registered in Malaysia (Kulim) and 1 at Derby (UK), implementation ongoing in Portugal, Thailand, India and Heywood
To increase the use of recycled packaging	Hand-packed cartons using recycled board increased from 29% to 41.5%	All hand-packed cartons made from recycled board	Investigations ongoing to find suitable board for machine packing	Ongoing	Unable to source suitable board for machine packing. Impact evaluation completed comparing recycled and virgin boards	**Objective closed.** Policy reversed, virgin board sourced from sustainable forestry operations used in preference. Impact minimized by light weighting	

To reduce and simplify packaging	Investigations begun to identify areas of concern to customers and community	No specific areas identified, targets to be determined for removal of composite and secondary packaging	Four potential product groups identified, investigations ongoing	A general reduction of 5% noted but no specific targets had been set. **Objective closed,** part of operational control to ensure legislative compliance and covered by remit for waste minimization	Audit guidelines updated to include annual packaging audits and identification of areas for improvement	Site-specific projects initiated in response to audit results	A general reduction in packaging noted (down from 7500 tonnes to 6000 tonnes)
To improve fuel consumption of car and commercial fleets	3 years of data reviewed, action plan defined	10.3% reduction achieved	Audit completed. Vauxhall Vectra identified as best environmental option	Vauxhall Vectra established as car of choice for Sales Team, 15% reduction achieved	Project refocused on fuel use for business travel as a whole. £100 000 invested in video conference facilities	Systems being developed to monitor use of video conference facilities and business travel	Unable to identify fuel savings from video conference use and determine a methodology to monitor business travel. **Objective closed,** refocused on process use of fuels covered by remit to improve energy efficiency

(*Continued*)

Objective	1995/6	1996/7	1997/8	1998/9	1999/2000	2000/1	2001/2
To ensure products and packaging safe for disposal	Guidance and expertise provided to NHS on key areas relating to waste management and incineration	Projects established to replace PVC in tablet blister packs and urine drainage bags. 5 products packed in polypropylene monoblisters and put on stability trials	Licence variations prepared for blister packs and Market acceptability being tested. Work ongoing to develop specifications for PVC-free urine drainage bags	Product licences varied but monoblisters not acceptable to the customer. Potential materials for PVC-free bags identified	Sample PVC free bags assembled. Project initiated to replace PVC used in blisters for condoms	Specifications for PVC-free bags approved. PVC replaced in condom blister packs. New project initiated to replace PVC used in Scholl blister packs	Continence care business sold. PVC replaced with A-PET in Scholl blister packs and in trays used for Bootle products
To identify and understand the environmental impacts of products and processes through a study of product life cycles .	Evaluations completed on tubular bandages and absorbent foam dressings. Policy established to purchase only yarns with OEKO-Tex certificates. Project initiated to	Evaluations completed on head lice/scabies treatments and urine drainage bags. Project initiated to replace use of PVC in urine drainage bags	Evaluations completed on anti-embolism products	Evaluations completed on compression hosiery	Evaluations completed on catheters, infection control and antacid products. EtO contractor influenced to initiated improvement project	Evaluation completed on surgical gloves	Evaluations completed on Scholl medicated mass products and condoms. Procedure under review – no longer useful in identifying improvement projects

	replace use of HFCs in polyurethane foam manufacture						
To replace all solvent based materials	**New objective**	Successful trials completed on in-house coating line, quotations being collated for new machinery	New machinery ordered	Solvent-based adhesive coating line replaced with a £80 000 hot melt adhesive line. Raised the profile of project to replace solvent-based adhesive used in Scholl medicated mass	Contractor completing EtO sterilization influenced to initiate project to fit abatement equipment. Licence application made to replace solvent-based adhesive in medicated mass	Abatement equipment fitted. New medicated mass supplied to Japan, clinical trials ongoing to provide data for UK licence submissions	Medicated mass project ongoing. Use of solvent-based inks in Malaysia reduced
To reduce wastage in the use of raw materials and other related sources		**New objective:** Funding from Groundwork Manchester and Environet 2000 accepted	Teams from Environet and ENERGI completed waste management and energy audits. Detailed project plans being	Environet project completed with projected savings of more than £100 000. Work being continued through Productivity	Sites set up improved waste management systems.	**Waste:** Oldham and Redruth completed successful projects to increased % waste recycled or reused. Spain reduced levels of all	**Waste:** Malaysia reduced rubber waste

(*Continued*)

Objective	1995/6	1996/7	1997/8	1998/9	1999/2000	2000/1	2001/2
			drawn up	Improvement Challenge at Oldham. Site specific projects being set up at all ISO 14001 sites		types of waste	
To minimize any potentially damaging releases to the aqueous environment				**New objective**	New effluent discharge system installed at Bootle and a new waste latex treatment plant installed in Spain	Waste water treatment plant upgraded in Malaysia	Key performance indicators for monitoring effluent being developed
To develop a methodical approach towards stakeholder dialogue and promote local activities						**New objective**	Oldham completed joint project with cardboard supplier, achieving savings of £65,000 pa, Malaysian sites completed successful joint projects with Shell and Isotron

Lex Vehicle Leasing Ltd

Written by Peter Knights, General Manager, Communications, and Helen Counsell, Quality, Environmental and Data Protection Manager, Lex Vehicle Leasing Ltd, Crossgates, Manchester and Marlow, Buckinghamshire, United Kingdom

Lex Vehicle Leasing Ltd (LVL) is a vehicle leasing company with around 600 employees at three locations within the UK. It therefore just fits inside the definition of an SME for the purposes of this book. Its business is in the leasing of company cars and vans to customers. This simple scope of 'leasing' covers a wide spectrum of service 'packages' offered to its broad range of customers from corporate multinationals to small businesses with 4000 vehicles on lease to personal contract purchases for individuals. These customers have diverse demands in line with their business activities and private needs as well as their financial resources. The number of vehicles leased at any one time is about 99 000.

Thus the business of Lex is in the provision of a service with no manufacturing assets or interests.

Introduction

Lex Vehicle Leasing Ltd is in the business of providing the consumer with one of the most environmentally unfriendly inventions of the twentieth century – the car. With this as the starting point it could be argued that they, Lex Vehicle Leasing Ltd, to demonstrate environmental responsibility, should close down and thus prevent 99 000 cars using the congested road network and polluting the atmosphere. A moment's reflection will show the futility of this approach. Another competitor or entrepreneur would merely increase their own business by another 99 000 cars to fulfil market needs. The consumer would still require a lease car. Another, less drastic, option would be for Lex Vehicle Leasing Ltd to dictate to its customers what sort of vehicle they should lease. Clearly insisting that a customer with few long journeys, mainly inner city commuting and no spouse or children *must* lease a small fuel efficient 'mini' car would be business suicide if that same customer is a managing director and needs a more opulent car for 'prestige' and business purposes.

Again, a moment's reflection would indicate that this client would merely migrate to another service provider. There would be no net advantage to the environment.

The two options above are tongue in cheek but nevertheless give rise to the very pertinent question: 'What can a car-leasing organization do to demonstrate it does have environmental responsibilities?'

Due to shareholder pressures, and of course general environmental press, as well as customers becoming more aware of environmental issues, Lex Vehicle Leasing Ltd began to seriously question how its activities could square up with seemingly contradictory business requirements – supplying customers with cars (bad polluters!) and yet behaving in a 'cleaner' way. A solution was found which appeared to satisfy this contradiction which was to supply the customer with the chosen car but give advice on the best way to use the car to minimize impact on the environment. Most customers would not object to this sort of advice. Some may heed it and some may choose to ignore it. Thus in late 1997, Lex set up an 'Environment Unit' with the brief of advising customers on environmental issues based upon best practices within the car leasing business, and to advise customers from an informed position.

However, it was soon realized that although it was advising customers of the most fuel-efficient vehicles and how best to improve fuel efficiency through better driving techniques, it had done little to change its own behaviour. Issues surrounding diesel versus petrol, route planning and timely vehicle maintenance were all well and good, but what had Lex Leasing Ltd done to show that it had got its own house in order?

ISO 14001 certification process

The Environment Unit looked searchingly inwards and decided the best way to address the issue was to demonstrate impartial evidence that Lex did indeed take its environmental issues seriously; this impartial evidence being in the form of ISO 14001 certification. With support from the Lex Board, members of the Environment Unit decided that external verification of its own environmental management processes was a must and that ISO 14001 certification was the method to demonstrate this.

The Environment Unit from the outset accepted certain criteria upon which Lex Vehicle Leasing programme could develop. This was called 'A Contract with the Future', an internal document which outlined the issues facing the company.

- That it (Lex Vehicle Leasing Ltd) was part of a global community from which it derives benefits and to which it owes a responsibility

- Its responsibility extended to lessening the impact of all aspects of its operations on that community and the environment on which it depends
- That the car – with its traditional technologies – has a detrimental impact upon the environment
- That CO_2 emissions from vehicles are a significant contributor to global warming
- That vehicle emissions have a significant negative impact on urban air quality public health and well-being
- In parallel with these thoughts of course it must not be forgotten that the car has brought great benefits to, and is an intrinsic part of, modern society.

How could Lex demonstrate to customers that concern for environmental issues was part of its culture, its daily habits and routines?

The preparatory review

The Environment Unit, with the assistance of external consultants, performed a rigorous review of all the Company's activities in late 1997 and came up with a set of environmental aspects. In essence, a PER was carried out. From this review a preliminary environmental statement was written in early 1998 with three main broad objectives of:

- encouraging the use of clean vehicles
- exploring new, sustainable areas of vehicle usage
- compliance with all legal requirements, reduction of pollution and commitment to continuous improvement whilst operating the business.

Aspects' identification

It was realized from the outset, that aspects would readily fall into two separate areas.

1 Aspects related to the provision on lease cars, at clients request and over which Lex Vehicle Leasing could not directly change i.e. only

advise. This represented the environmental impact of some 99 000 vehicles and 500 customers, i.e. so-called indirect aspects.

2 Aspects relating to Lex's own internal operations: and which could be directly influenced by senior management setting policies, decision-making as well as communication and training (essentially an EMS), i.e. the direct aspects.

It was also realized that *all* indirect aspects would have resultant significant impacts and thus during the evaluation process for significance a different approach was needed.

Aspects' evaluation for significance

Because of the large numbers of vehicles on lease, the indirect aspects identified far outweighed the direct aspects in terms of significance. However, these indirect aspects are related to the objective of influencing people's perceptions and trying to influence their purchasing decisions, their driving habits and their whole approach to use of vehicles. In short, persuading customers to change their life style. No trivial task and one which would take a long time to achieve!

For example, Lex was responsible for a significant number of cars on the roads and with that amount of purchasing power some level of influence could be made upon car manufacturers. This need not be dramatic but once it was known throughout the motor industry that yet another large purchaser of cars was placing more emphasis on green issues in its purchasing philosophy, then this must have a positive effect on car manufacturers to step up their research programmes into cleaner more efficient vehicles.

The order of significance then became one of prioritization as the issues – mindful that some of the *more insignificant direct impacts* were going to be achieved ahead of the more *significant indirect impacts*.

Management programme

The three broad objectives stated within the environmental policy were now formalized and an element of focus was achieved by drawing up a management programme. This is shown in summary in Figure 5.1 and for each of the three objectives, several targets (and/or time-scales) were

OBJECTIVES	TARGETS	MEANS
1) ENCOURAGE THE USE OF CLEAN VEHICLES	i) A commitment to reduce the CO_2 emissions from our staff car fleet by 20% by the year 2010	Introduction of Lex Company Car Policy
	ii) Encourage customers and other stakeholders to operate their fleets of vehicles in ways that mitigate environmental problems	Education through the 'Breathe Easy Campaign'
	iii) To incorporate environmental performance as part of the purchasing policy	Undertake a review of the Company's procurement criteria
	iv) Encourage utilities and gas supply companies to provide a ready supply of gas or on-site electric recharging facilities	Meetings with representatives from gas and electricity utilities, as well as vehicle conversion specialists to facilitate collaboration
2) EXPLORE NEW AREAS OF VEHICLE USAGE	v) Investigation of opportunities for moving goods and people in more environmentally responsible ways	1) Use of video conferencing 2) Use of business travel diary
	vi) Seek ways in which to reduce the use of private vehicles commuting to Lex offices	1) Encourage use of bicycles 2) Public transport schemes/park and ride
3) COMPLIANCE WITH ALL LEGAL REQUIREMENTS, REDUCTION OF POLLUTION AND COMMITMENT TO CONTINUOUS IMPROVEMENT WHILST OPERATING THE BUSINESS	vii) to set annual targets for reduction in energy use in comparison with business activity	1) Convert lighting 2) Poster campaign to raise awareness
	iii) set annual targets for the reduction in paper use and increase in paper that is recycled	1) Use of e-mails rather than fax 2) No printing out of e-mails 3) Poster campaign to raise awareness

Figure 5.1 Lex Vehicle Leasing Ltd management programme 2000–10

established. At this stage not all of the targets were either identified or the means of achieving them fully finalized.

Note: *The Company adopted a definition so that 'clean' would not be so open in its meaning.*

Clean means reduced exhaust emissions. Lex adopted the California Air Resources Board's criteria of Low Emission Vehicles (LEV), Ultra Low Emission Vehicles (ULEV) and Zero Emission Vehicles (ZEV). These lower pollution levels are achieved in two ways – by designing vehicles to use less fuel, and/or by using alternative propellants such as electric or gas.

Targets and means – further explanations

Target i) A commitment to reduce the CO_2 emissions from our staff car fleet by 20% by the year 2010

Means: The phased introduction of the Lex Company Car Policy. This involved elements of giving cash incentives to staff to accept smaller fuel-efficient cars instead of the larger car that their status in the Company warranted.

Target ii) To encourage customers and other stakeholders to operate their fleets of vehicles in ways that mitigate environmental problems

Means: Education aimed at two separate groups:

a) Education of staff:

It was recognized that for Lex to advise stakeholders, they themselves required training. This was to be through briefings and sales training sessions. This included the distribution of the environmental policy and ensuring that all senior management were well versed in the environmental objectives.

Clean driving training was also offered to staff. Additionally, an alternative fuels reference file and pricing guide was put together for the sales team, to enable them to provide the best available to clients. This information included manufacturers' information on models and their CO_2 emissions; Government information such as company car taxation and legislation; trade association information; grants available for conversion from petrol to LPG.

b) Education of stakeholders:

Using the 'Breathe Easy Campaign'

This was an awareness promotion to increase stakeholders' awareness in how they could operate their fleets or vehicles in ways that mitigate environmental impacts. This included consideration of new fuels and technologies, improved vehicle maintenance (such as the importance of checking tyre pressures and regular servicing to improve fuel consumption); the importance of driving habits (i.e. defensive driving).

The 'Breathe Easy Brochure' was published and given to clients with all new hired cars. This explained issues surrounding vehicle usage and the environmental and public health impacts of exhaust emissions.

1) Journey planning to reduce wastage of fuel by planning the most economic (usually the shortest) route – introduce clients to the most modern system of journey planning – essential software-based programs.
2) Clean vehicle guide – published by LVL in conjunction with manufacturers and relevant associations.
3) Operation of a telephone helpline

Additionally, Clean vehicle demonstration days were also offered as another means of getting the message across to both staff and customers.

Target iii) To undertake a complete review of the company's procurement criteria with the aim of incorporating environmental performance as part of the policy. The focus of this target was a complete review of the company's purchasing criteria with the ultimate target of incorporating environmental performance as part of the procurement policy. As an owner of some 99 000 vehicles, which were all purchased new from manufacturers (albeit via dealership chain) then some influence should be possible.

Means: To develop a set of criteria, which, if all other factors were equal, would give preference to a supplier who is able to demonstrate the best environmental credentials. Such criteria would include whether environmental policies and procedures were in place or ideally, certification to ISO 14001 had been achieved.

Target iv)	To encourage utilities and gas supply companies to provide a ready supply of gas or on-site electric recharging facilities for alternative fueled vehicles. One of the major challenges for alternative fuelled vehicles is the lack of infrastructure for refuelling compared to the existing infrastructure which exists in all industrialized countries for petrol and diesel fuelled vehicles. Lex decided to stimulate interest with both vehicle converting companies and some major oil companies and electricity utility companies. Again this was seen as a longer-term target and measurable by the increasing uptake of converted vehicles by customers and increasing numbers of refuelling locations.
Means:	Meetings with representatives from gas companies and electricity utilities as well as vehicle conversion specialists to investigate the potential for collaboration.
Target v)	The investigation of opportunities for moving goods and people in more environmentally responsible ways.
Means:	This included the use of video conferencing to reduce the amount of fuel, time lost travelling and pollution when staff needed to have meetings with staff from other offices. Additionally, a business travel diary was piloted to see if this could facilitate at the very least car sharing for business purposes.
Target vi)	To reduce the impact of staff travel arrangements commuting to our offices, encourage staff to travel in a more environmentally responsible way and monitor the effectiveness of the actions taken.
Means:	Determine current travel arrangements. Trial schemes with local authorities and private providers of park and ride, taxi rounds etc. services. The provision of showers, changing rooms and secure cycle stands to encourage staff to cycle to the workplace. The provision of public transport timetables in staff canteen and other areas to encourage interest in public transport.

Target vii) The setting of annual targets for reduction in energy use in comparison with business activity.

Means: Electricity, water and gas consumption monitored and data analysed. Convert lighting systems within the sales office to more efficient low-energy type. Conduct awareness campaign to switch off all printers, PCs, fax machines and photocopiers over nights and weekends. Investigate ways of auditing success of this. Campaign to educate staff not to open windows when air conditioning is switched on. Investigate the use of separate thermostats for office heating systems. Targets could be set for reduction in consumption.

Target viii) The setting of annual quantifiable targets for the reduction in paper use and an increase in the proportion of used paper that is recycled.

Means: Not printing out e-mails; enable PCs to fax information directly rather than printing out, then faxing; modify printers to default to double side printing; poster campaign to raise awareness of paper consumption; investigate potential for reuse of paper for photocopiers or rough pads.

Successes and failures in the achievement of these objectives

The numbers of dual fuel vehicles on the fleet of 99 000 vehicles has risen to 1200 over a period of some 4 years. Internally, for staff commuting a Car Share Programme is now in operation and video conferencing as a means of reducing travel to other sites has grown.

Measurable achievements include video conferencing: savings of the equivalent of 700 000 miles of driving, 200 tonnes of CO_2 emissions and 401 working weeks (by reducing the loss of management time spent driving).

Waste Recycling Group plc

Written by Andy Harris, Group Safety, Health and Environmental Manager, Waste Recycling Group Plc, Doncaster, South Yorkshire, United Kingdom

The past, present and future of environmental management systems within WRG plc are described and demonstrate that not only does more demanding environmental legislation within this highly regulated industry drive continuous improvement, but as the management system has matured and with the passage of time, the needs of other interested parties has emerged. These interested parties have needs which must be addressed thus perpetuating the continuous improvement cycle.

The past

WRG plc was formed out of amalgamation of several smaller waste management companies in the mid to late 1990s. Operations include transfer stations, landfill sites, household waste recycling centres, composting, waste to energy and energy from waste plants, treatment plants and a quarrying business.

The main 'customers' are generally local authorities whose domestic and commercial waste they accept, manage, recycle, process and dispose of. The following table indicates the diversity of activities as at 2002:

Operational landfill sites	62
Closed landfill sites	21
Transfer stations/recycling centers	21
Household waste sites	57
Compost facilities	21
Liquid waste treatment plants	3
Energy from waste plant	1
Quarries	6
Total number of licensed sites	192
Number of employees (average for year – full-time equivalents)	861

Because this amalgamation of these quite diverse smaller companies brought different management systems, structures, policies and procedures it was decided by top management, in the interests of minimizing any potential for environmental harm, that a coherent safety, health and environmental management system was required. This would enable waste to be disposed of in a safe, effective and economic manner. The Company was well aware that failure to comply with legislation would result in fines, licences being revoked, potentially high costs for remediation (under the 'polluter pays principle'), and damage to the organization's image.

The first step was to achieve ISO 14001 for the following reasons:

- to visibly demonstrate environmental credibility – certification to an International Standard
- to improve environmental performance
- to facilitate and enable the newer acquired companies to integrate their management systems
- to standardize WRG plc's approach to environmental management
- to improve the financial position of the organization through better risk management

A Safety, Health and Environmental (SHE) Manager was appointed in order to drive the process forward. The time-scale for the implementation of ISO 14001 was shortened, and the complexity of the work lessened when it was apparent that many of the requirements of ISO 14001 were already implemented, due to the legislation framework for the waste management industry.

Thus elements of the requirements of ISO 14001 such as environmental aspects, register of legal requirements, continuous improvement, operational controls, emergency procedures, training and competence, control of documentation and records were all described within the necessary planning permissions, especially for landfill sites, and site licences (with their attached site working plans).

Three sites were chosen to pilot the implementation of ISO 14001 so that any learning experiences could be applied when rolling out the system across the whole Company. These pilot sites were existing landfill sites – such sites chosen because of their diverse activities and strict enforcement of regulatory compliance. Additionally, an overriding objective in terms of this early development was to have a system that would become an integral part of the day-to-day operations and management at each WRG plc site. Measures of improvements in SHE performance and increased awareness of individuals was sought rather than just obedience to a system and a 'paper chase' perception of ISO 14001.

Figure 5.2 shows the then 1999 management programme with most activities concerned with implementation, 'roll out' target dates and an overall aim of ensuring legal compliance.

Implementation of ISO 14001

The implementation process was perhaps no different from any other ISO 14001 implementation programme, but there were key areas which were different and unique to the waste management business because of the existing regulatory framework and enforcement regimes. Clearly, more time was spent in developing and facilitation at those sites associated with high sensitivity in the public's perception and the media's attention. Landfill sites required more diverse and critical operational controls than a household waste site for example. Two key points to note during implementation were:

1) Identification of environmental aspects

The most significant aspects are invariably to be found on landfill sites where the following were identified. (Such aspects are found on all landfill sites and are well documented, researched and well understood within the waste management industry and the regulatory bodies alike):

- potential for contamination of groundwater
- potential for pollution of surface waters
- potential for pollution – escape of leachate
- potential for nuisance and danger of landfill gas migration
- potential for nuisance – generation of odours

By the very nature of the landfill process, these aspects are inherent and cannot be eliminated by current technology. Therefore operational controls need to be robust to reduce such potential to pollute or cause nuisance.

	SUMMARY OF 1999	ENVIRONMENTAL	PLAN
No.	**Objectives**	**Key stages**	**Target dates**
1	Obtain ISO 14001 for 3 sites by December 1999	1) Gowy landfill site 2) Buckden landfill site 3) Welbeck landfill site	Sep-99 Oct-99 Nov-99
2	Roll out the management system to aim for all sites to be registered by end 2001	Issue to sites: 1) Health and safety manual 2) Policy and procedures 3) Environmental and quality assurance procedures	Jun-99 Aug-99 Sep-99
3	Develop leachate and gas management plans for each site by mid-2000	1) Identify significant environmental aspects 2) Prioritize by site 3) Develop plans by priorities	Jun-99 Aug-99 Aug-00
4	Establish an accurate database for all recycling activities across the whole group by August 2000 to enable measurement of year on year improvements (using 1999 as the baseline year)	1) Gather data 2) Set up databases 3) Formulate a measurement system to give 'normalized' results 4) Produce report	Jul-99 Nov-99 Dec-99 Dec-99
5	Develop and commission 2 waste to energy projects in 1999	1) At Buckden landfill site 2) At Boston landfill site	Sep-99 Sep-99
6	SHE manager audit all sites in 1999	Ongoing	Dec-99
7	Perform regulation 15 risk assessments	1) Gainsborough 2) Skelbrooke 3) Long lane 4) Leadenham	Sep-99
8	At least 3 sites to demonstrate evidence of communication with local community	1) Database of sites to be set up 2) Liaison plan to be drawn up	Aug-99 Sep-99
9	All sites to have minimum COTC* cover as defined by the Environment Agency A minimum of 1 person trained for first aid (* Certificate of Technical Competence)	1) Check minimum required 2) Decide on COTC training 3) Achieve minimum cover 4) Train for first aid	 Aug-99 Aug-00
10	Set up supplier database – for assessment of environmental responsibility	1) Business critical suppliers 2) Devise scoring system 3) Send out questionnaires	Aug-99 Sep-99 Oct-99

Figure 5.2 WRG plc summary of 1999 environmental plan

Additionally, the following aspects were found to be less significant and mainly depended on the location of the site, be it a transfer station, landfill site or household waste site. They were identified either through discussions and feedback from local residents, members of the public or the local regulatory authority:

Aspect	Cause
Generation of dust	Deposit of waste in dry weather
Noise	i) the waste itself (glass, metal)
	ii) traffic – both members of the public and WRG plc
Litter	Lighter waste blown by the wind
Visual impact	Waste is not visually attractive
Use of water resource	For damping down and cleaning public access
Spillages of fuels or other liquids	Fuel from WRG plc vehicles such as bulldozers, compactors, heavy goods vehicles
Suppliers or contractors using sites	Ensuring such contractors act environmentally responsible

2) Documented system/Operational controls

Especially on landfill sites, the requirements of the regulatory framework meant that most of the documented system existed but required some additional structure and controls putting in place. The following legal documents formed the core of the management system and required operational controls to demonstrate compliance:

Permit (site licence)

The specific set of conditions that govern the operation of the site. Failure to comply with these is a criminal offence under the Environmental Protection Act and or the Pollution Prevention and Control Act.

Working plan
The Company's statement on how they are going to operate the site in compliance with the conditions imposed on the licence.

Regulation 15 risk assessment
The regulation 15 risk assessment is a study and objective assessment of the main environmental risks to groundwater posed by a site.

Planning permission
This document sets out the conditions to be adhered to for this specific planning approval. It is usually accompanied by the planning application, which includes all the initial environmental risk assessments. It may also be accompanied by a legal agreement to complete a specific action usually linked to restoration of the site.

Related legal documents
Other documents that exist and could be:

- Discharge consents to surface waters
- Discharge consents to foul sewer
- Lease agreements

The present

Some 5 years on from then, the WRG Group has continued to expand, and now has a large number of sites spread throughout the UK. This expansion has resulted in additional challenges but they can be faced with confidence because of the maturity and strength of the management system. These challenges include:

- Increased number of enforcement notices from the Environment Agency
- Submission of landfill conditioning plans well before deadline set by the Landfill Directive
- Rapid response for systems to store and transport fridges to aid local authorities in complying with the Ozone Depleting Substances Regulation (ODS).

The management system itself has continued to develop from a 'stand alone' ISO 14001 EMS and is now a fully integrated management system including a host of policies:

Policy	Enactment
Health and Safety	Risk assessments
Environmental	Risk assessments
Social	Human rights
Ethical	Bribes/gifts etc.
Policy in disclosure (whistle blowing)	Protection of employees who flag up malpractices
Community relations	Interaction with stakeholders
Training	All employees will have access to improving their skills
Equal opportunities	No discrimination for advancement

And the system is designed to ensure:

- That all legal requirements, regulations and standards are complied with.
- That WRG plc works to improve standards, increase environmental awareness, improve customer relations and service levels and commitment amongst staff. Achieved through training and encouraging the adoption of sound environmental principles amongst contractors, suppliers and customers alike.
- That WRG plc respond positively to all quality and environmental developments and review such issues with the appropriate authorities, local community and other bodies.

This is reflected in the WRG plc environmental policy (see Figure 5.3).

Note: This policy replaced an earlier one, which was valid and appropriate, but its scope was only in terms of the requirements of ISO 14001.

The future

Three activities are described below, which while ongoing, indicate where future management input will be required as society's needs and requirements from the waste management and recycling business become more sophisticated. Of note is that communications were recognized by WRG plc

Waste Recycling Group plc is committed to achieving high performance throughout the business. Compliance with all environmental legislation pertinent to our activities is a minimum requirement and an integral part of this policy. In addition Waste Recycling will:

- Pursue continuous improvements in environmental performance and the management system.
- Develop and maintain activities to protect and enhance the environment and prevent pollution.
- Promote waste recycling and recovery and endeavour to replace the use of non-sustainable natural resources.
- Identify environmental risks and use all practical measures to reduce risk.
- Have an ongoing commitment to informing and educating legitimate interested parties about our activities.
- Have an ongoing commitment to develop all personnel.
- Wherever possible source materials and services locally to minimize transport impacts and support the local economy
- Work with suppliers to minimize the impact of their operations on the environment.

The ultimate responsibility for environmental performance lies with the Chief Executive who will ensure that it is given equal priority with other major business objectives. Implementation of this policy is a line management responsibility at all levels together with participation of all employees. Specific arrangements and organizational responsibilities are detailed in the policy and procedures manual (volume 1). Staff are reminded that adherence to this policy is a condition of employment. The Board will review the policy and its implementation annually and publish the results of the review. It will be revised and updated as necessary by the Group SHE Manager.

Figure 5.3 WRG plc environmental policy

as being a key to future operation of the business (as required by ISO 14001) and this has developed and certainly proved its worth. It has been possible to go beyond just compliance with legislation, and WRG plc can readily demonstrate further examples of continuous improvement for the future. The three activities also demonstrate that environmental improvement cannot always just be viewed in isolation. Customers (stakeholders) are becoming more demanding in their requirements and this is likely to drive further improvements, and not just environmental.

1) Stakeholder engagement

Certainly, the rise of the power of the stakeholder has been identified and recognized as a force to be met and the methods of communication with such stakeholders documented to demonstrate how WRG plc engages at all levels of society. Some of the day-to-day engagement is regulatory – for example, planning and environmental – but the table below shows activities that go beyond the statutory requirements.

Stakeholder	Description	Methods of communication
Shareholders	Waste Recycling's main shareholder, Kelda Group plc hold a majority share holding and have two non-executive directors on Waste Recycling's Executive Board	• Annual Reports • Interim statements/circulars • AGM • Website • E-mail • Telephone • Correspondence Two-way communication
Customers	Local authority customers provide nearly 50% of business	• Contract meetings • Site inspections • H&S inspections • Local manager communication • Website • Newsletter Two-way communication
	Trade waste customers	• Sales visits • Duty of care site visits to site • Local manager contact • Website • Newsletter Two-way communication
Employees	All staff, hourly paid and part-time workers, including those on temporary contracts	• Training programmes • Appraisal process • Site meetings • Site notices • Safety committees • Newsletters Two-way communications
Local communities	Representatives of local communities. Liaison committees include other stakeholders i.e. Environment	• Open door policy • Local manager contact • Liaison meetings • Website • Individual and group visits to site

Stakeholder	Description	Methods of communication
	Agency and authorities	• Exhibitions • Newsletters • Public meetings • Media statements and company announcements • Complaints monitored Two-way communication
Environment Agency	Main environmental regulators	• Site inspections • Site meetings • Regional meetings • Correspondence • Seminars and training Two-way communication
Health and Safety Executive	Main health and safety regulator	• Site inspections • Site meetings • Regional meetings • Correspondence • Seminars Two-way communication
Planning authority	Local authority planning dept	• New sites • Existing sites for modifications • Local liaison committees Two-way communication when needed
Environmental Health	Local authority environmental health and health and safety regulator	• Only when needed Ad hoc communication
Contractors	On-site contractors providing engineering works, electrical works, litter picking etc.	• Local manager contact • Small contractors control • Permits where required • Top contractors monitored by SHE dept Two-way communication

Stakeholder	Description	Methods of communication
Engineering partner	Main civil engineering contractor	• Operations meetings • Contract meetings • Audits Two-way communication
Monitoring services	Main environmental monitoring and lab services	• Environmental manager reports • Local manager contact with monitoring staff Two-way communication
Landfill gas contractors	Provision of power generation and landfill gas management	• Gas management team • Local manager contact • Site meetings Two-way communication
Other suppliers	Consumables, PPE and some national purchasing contracts	• Local manager contact • National contracts managed centrally Two-way communication

Such engagement has been extremely useful and highlighted issues which if not addressed in the future could give rise to regulatory intervention. Generally, this has fed through to objectives and targets being set at local levels i.e. unique to a particular type of site in a particular location.

2) Environmental reporting

Customers and other key stakeholders now require WRG to report on issues ranging from the environment to human rights, giving a snap shot of the Company's all round health during the year and showing how the Group has developed and improved. They also require demonstration of commitment and evidence that systems are embedded in the culture and day-to-day operation of the Group. This means the development of benchmarks or key performance indicators to show how these systems are implemented and how improvements are measured. Our indicators are specific to the waste management sector and the work of the Group and are selected with care to give a genuine overview of our performance.

WRG plc has compared a number of current corporate responsibility reports and referenced various standards and reporting protocols, including the Global Reporting Initiative (GRI) and the Social Accountability 8000 (SA 8000) standard. We have also been involved in the work carried out by the Green Alliance in setting environmental indicators for the waste industry; the Group has signed up to the published stage 1 indicators and is well on the way to compliance with the more challenging stage 2 indicators.

3) Verification

There is increasing pressure for all corporate responsibility reports to be externally verified or to undergo some form of third-party scrutiny. Currently, WRG plc is independently assessed or examined in the following areas:

Area	Body	Scope	Details
Health and Safety	Internal auditors	Internal Audit – Evaluate Group policy, record systems in operation and conduct compliance testing as appropriate; provide assurance over the adequacy and effectiveness of internal controls applied within the health and safety system	Risk management including • Management system • Accident stats • Procedures on site • SHE audits • Management review
Environmental Management System	SGS UK Ltd External certification body	ISO 14001 Environmental management systems on landfill, household waste sites, transfer stations, compost sites and energy from waste overs systems, objectives and targets	Environmental management system • Procedures • Objectives and targets • Policy About 40 individual sites sampled per year

Area	Body	Scope	Details
Environmental	Environment Agency site inspections	Legislation – site licensing, operator pollution monitoring appraisal. Environmental monitoring data	Licence conditions covering environmental performance, duty of care, operator scores, monitoring and engineering standards. Covers all licensed sites
Risk management	Heath Lambert, insurance brokers	Health, safety and environmental risks on site	Health and safety and environmental activity at site level. Six sites sampled per year selected from higher risk sites
Duty of care, some health and safety	Various customers	Customer activities	Duty of care compliance. Safe working on site
Health and Safety	Local Authority contracts	Health and safety on some sites covered by Local Authority	Health and Safety • Traffic management • Safe working
Corporate Social Responsibility (CSR)	Investment funds, research organizations	All areas of CSR within the Group. Both questionnaires and site visits used for verification	• Environmental • Health and Safety • Human rights • Ethics • Community

Summary

As WRG plc continues to grow, ISO 14001 becomes only part of the wider requirement to report and be judged on a whole raft of corporate issues.

New ideas, concepts and information on corporate responsibility reporting will continue to be taken on board and the Group will be under pressure to produce a report that is relevant, and can be accessible to all stakeholders through the use of openness, accurate data and straightforward language.

Chapter 6

The auditor and auditing standards

Introduction

ISO 14001 was designed to be an auditable standard and, within this book, references have been made to 'certification bodies' and 'third-party' auditors who will undertake this auditing process. Having committed themselves to obtaining ISO 14001 certification, organizations naturally enough focus on gathering environmental information, allocating resources and employing environmental consultants (if required). Budgets will have been prepared, and personnel coached in how to answer the auditor's questions. A tremendous amount of preparation will have been done to enable certification to be obtained within the planned time-scale. However, it could be that no face-to-face contact has been made with the representatives of the certification body (that is, the auditors) until implementation is well in progress and the deadline for the stage 1 audit is fast approaching.

Although some information is available from certification bodies setting out the assessment and certification process (see Chapter 3), little or no information is readily available about the actual auditor, or team of auditors, who will visit the organization's site to perform the audit. This individual, or team, will need access to all departments, will disrupt normal organizational activities, demand valuable management time, interview (possibly very nervous) staff, and will have the power to recommend certification, or otherwise to the organization. So it is wise for an organization to do just a bit more preparation and to find out what sort of individual (or individuals) will actually arrive on the first day of the audit. Such preparation can only be a good investment of time so that a good rapport can be established with the auditors as quickly as possible. A good relationship will be essential between organization and auditors so that:

- Disruption to the business is minimal
- No differences of opinion or interpretation of the Standard leads to animosity
- Personnel do not feel threatened or intimidated
- The audit proceeds smoothly

The auditor, or auditors, will have to perform a physical 'walkabout' of the site, ask questions, take notes, and make as objective a judgement as possible on the sample of the environmental management system that is presented in evidence. Advances in video techniques and interactive information technology may one day change this but, for the foreseeable future, this appears to be the only (if somewhat imperfect) method of verification and relies heavily upon the concept of a 'standardized auditor'.

A standardized auditor cannot exist, because being human, the auditor like everybody else has a unique set of characteristics (including fallibility). However, some of the processes used by certification bodies to 'calibrate' auditors will be described. This calibration ensures that a minimum level of consistency of auditing is achieved.

Auditor characteristics

Organizations implementing ISO 14001 may have been certified to ISO 9001:2000 (or previous ISO 9000 series) for some years and will be familiar with being visited by a quality assurance auditor on scheduled surveillance visits.

Inevitably, because of quality assurance's long history, it has received some criticism in terms of delivering improvements to certified organizations and some myths have grown up, and been perpetuated – some unsubstantiated; others, unfortunately, having some substance. Certification body auditors have received some of this criticism with stories of auditors' conduct ranging between the superficial disinterested style to the arrogant 'nit picking' style.

In reality, auditor behaviour is somewhere in between these two extreme stereotypes. Auditors have to be reasonably confident in what they are doing and should present themselves to the organization as assertive. Unfortunately, in the stressful situation of an audit, this can be misinterpreted as arrogance. Almost every other day, auditors are facing new organizations – complete strangers to them – and they want to get the best out of the person being interviewed as quickly as possible. Upsetting people will create antagonism towards the auditor. The willingness to answer the questions honestly and in an objective and helpful manner will be lost. Personnel are often quite defensive when being interviewed about their tasks so putting them at ease is an essential requirement. An auditor, therefore, needs to be able to relate to all sorts of individuals and also be a good listener. Auditors need excellent interpersonal skills and need to establish a good working relationship with the organization very quickly. Time is usually against them – due to commercial considerations – and so, during the audit, 'small talk' may be at a minimum.

Auditors, like all employed personnel, are selected via an interviewing process in which interpersonal skills are assessed as well as the more tangible qualities of training, education and experience. Auditors, with these skills and personal qualities, are looking for the level of compliance with the Standard, and not the extent of non-compliance. Auditors also want the audit to assist the organization in meeting its business objectives – to add so-called 'added value' to the audit. To focus on purely negative issues and raise corrective action requests does not give job satisfaction. The auditor would like to feel that value has been added to the organization's system, through the mechanism of opportunities for improvement to either improve or simplify the system, and yet still meet the requirements of the Standard.

The organization, rightly or wrongly, expects the auditor to know their (the organization's) particular business immediately. In an ideal world, the auditor could spend several weeks at the organization's premises, learning the processes, getting to know personnel and their responsibilities as well as assimilating the unique language and culture that any business possesses. Unfortunately, no organization is willing to pay for this so the auditor must be able to assimilate this information very quickly, on-site, as well as performing the audit effectively.

Auditor qualifications

As previously suggested in Chapter 3, the stage 2 audit may require an auditing team to ensure that the appropriate skills and experience to perform the audit are present.

The team approach needs some further explanation. The required expertise to conduct a fair and valid assessment may not rest in one individual. The individual who has expertise of all national and international legislation, and has had practical industry experience in all industrial sectors, and is also an experienced environmental auditor probably does not exist. The years of experience and training, the tremendous amount of information to be assimilated, as well as the requirement to be continuously up to date on the many related environmental issues, would place unrealistic demands on any one individual. Therefore the team will comprise of individuals who, collectively, will bring the correct expertise to the assessment. They will be led by one of the team – designated the team leader.

However, it is understood by all interested parties, notably the certification and accreditation bodies, that such individuals will broaden their expertise (that is, acquire new skills via briefings from experts, specific training and participating in audits with technical experts). For example, an EMS auditor, with a scientific education, training and industry experience, will readily assimilate knowledge of forestry practices by private study and, of more importance, by being on stage 1 and stage 2 audits with forestry experts. Such an auditor will be able to put such specific forestry practices into the context of the (generic) Standard, and perform a valid audit of the forestry organization's EMS. This does not of course create a new forestry expert but it creates an EMS auditor who, in the future, will be able to perform valid audits, with a positive approach to seeking improvements to forestry management environmental systems generally, by virtue of his experience.

The qualifications and skills required within the auditing team are four-fold:

i) Management system auditing capability

ii) Environmental competence – including regulatory and legal compliance

iii) In-depth knowledge of the Standard – ISO 14001

iv) Technical knowledge of the industry

These qualifications are described below.

i) Management system auditing capability

In general, most environmental system auditors will have gained auditing experience through auditing of ISO 9001 quality assurance systems. This will have given them experience of knowing how to phrase questions to get the best answers from personnel being interviewed. They will be familiar with the structure, composition and hierarchy of documentation of a management system and, as discussed in Chapter 4, there is now alignment in the philosophy and structure between ISO 9001, ISO 14001 and OHSAS 18001.

It is also likely that they will have a wide experience of auditing to ISO 9001 organizations with activities relevant to environmental control. Such activities could include the operating of:

- Landfill sites
- Waste management and disposal
- Waste incinerator plants
- Water treatment plants
- Effluent plants

This background and experience will provide a good foundation for ISO 14001 auditing skills.

ii) Environmental competence

Environmental competence means the possession of sufficient experience and knowledge to identify readily an environmental aspect and to judge its likely level of significance. Being aware of, and understanding the current techniques for control and mitigation of such an aspect is also a measure of auditor competence. When judging the order of significance, therefore, the auditor will consider all the aspects of:

- Atmospheric emissions
- Discharges to aquatic environment
- Waste management and disposal
- Contamination of land
- Effects on ecosystems
- Nuisance pollution

The auditor must also have a clear understanding of the indirect aspects of the organization being audited. Indirect environmental aspects are the aspects particular customers and suppliers of the organization being audited are likely to have.

Auditors themselves will possess relevant professional and technical qualifications such as botany, biochemistry, chemistry, forestry management, etc. and will have had structured training in environmental issues.

It is also clear that a good knowledge of pertinent environmental issues and legislation is necessary so that the auditor can judge whether the system being audited will deliver performance improvement as well as regulatory compliance. The need to exercise such environmental judgement, or competence, is proportional to the environmental profile of the situation. High-level competence is often required in the chemicals or power generating sector, for example, where the potential for environmental incidents are higher, environmental issues are broader and far reaching and tighter legislative controls are in place.

ISO 19011

At this point, it is timely to briefly review ISO 19011, the 'Standard' for auditing, which addresses in particular this subject of competence.

The title of ISO 19011:2001 is 'Guidelines on Quality and Environmental Management Systems Auditing'. It was written by a joint ISO/TC committee in response to requests to replace the separate quality auditing standards (ISO 10011-1; ISO 10011-2 and ISO 10011-3) and environmental auditing standards (ISO 14010, ISO 14011 and ISO 14012).

Within ISO 19011, in section 7, there are details of the required education, audit experience, auditor training, work experience as well as personal attributes with sets of tables listing recommendations for the above, for auditors. There are linkages here to Accreditation criteria (see also Appendix II) which places much emphasis on the certification bodies operating robust systems and processes to assess the capability and ongoing competence of auditors.

Specific required competencies are suggested such as:

- Environmental management principles and techniques – such as understanding environmental terminology; environmental management tools (such as life cycle assessment)
- Environmental science and technology – such as understanding the impact of human activities, environmental media; interaction of ecosystems; general methods of environmental protection; technologies to prevent pollution; monitoring and measuring techniques

And although minimum-level academic qualifications are suggested in ISO 19011, the above abilities and skills do seem to suggest that most environmental auditors will possess a much higher level of qualification – possibly up to a post-graduate degree in an environmental discipline.

iii) In-depth knowledge of the Standard

Obviously, the auditor needs to know ISO 14001 intimately – this is a fundamental requirement for effective auditing. Such an in-depth knowledge can only be gained by studying the Standard and applying that knowledge in practical auditing situations. Reading of the Standard can of course be done in isolation but auditors will have also participated in workshops with colleagues to discuss areas of interpretation. Interpretation in this sense refers to words, phrases, sub-clauses or clauses within the Standard that could suggest different meanings to different auditors.

iv) Technical knowledge of the industry

The 'language' of the industrial sector being audited needs to be understood by the auditor. Knowledge of current best practice, with regard to environmental control within a particular sector, helps to ensure a meaningful audit. The need for such specific technical knowledge is also proportional to that sector's environmental profile, which can change over time depending upon new research or new media interest and public concern (for instance disposal of waste to landfill, waste dumped at sea and incineration of waste have all had periods of increased focus for regulatory bodies and stakeholders). Such knowledge will have been gained through previous operating experience within that industry. Most environmental auditors will have had previous employment experience though not necessarily in an audit-related function.

In order to satisfy accreditation criteria, a member of the auditing team who can offer such industry experience must be present at the stage 2 audit. Clearly, this industry experience must be reasonably substantive. What is required is a level of expertise gained by operating at a technical or managerial level for a reasonable period of time so as to assimilate the 'norms' and working practices of that business sector.

The technical expertise requirement for a team composition is carefully controlled by the certification bodies. This level of expertise requirement is carefully monitored by the national accredited body. Auditors are required to provide documentary evidence – which may be scrutinized by the accreditation body – as to why they as an individual, and as a team, are capable of performing a fair and valid assessment for each particular organization. Such documented evidence addresses the previous requirements (i) to (iv); it also acts as a safeguard for the organization being audited, ensuring that they are getting auditors who as a team will be aware of the norms applicable to their industry. This evidence also ensures that eventual certification is meaningful, credible and will withstand scrutiny from stakeholders.

To sum up this section, third-party auditors require a high level of education, training and experience in order to perform their jobs effectively. Organizations, themselves, should enquire as to the environmental background of the auditors, although auditors should be pro-active in this respect and tell the organization something about themselves prior to the audit. The opening meeting is an ideal opportunity for such enquiries. However, it does tend to be the certification bodies who set minimum requirements – as they have to be mindful of the fact that they are audited by accreditation bodies and could be taken to task if their auditors do not demonstrate a certain minimum level of competence in line with ISO 19011 and accreditation criteria.

Auditor methodology

Going back to the theme developed in the section describing auditor characteristics, there has been, it must be said, a history of emphasis on the structure of system documentation for quality assurance, from both consultants designing the system, and certification bodies auditing such systems. The ability of the system itself to deliver true quality and customer satisfaction, was quite often lost in the midst of documentation. Time has moved on and with ISO 9001:2000 now more able to assist organizations deliver true improvements, such bureaucratic systems are becoming rarer. Environmental auditors now seek out whether the system is based upon the management of environment aspects, delivery of environmental performance and improvement and delivery of regulatory compliance. The decision-making processes, how environmental aspects are given a measure of significance, and whether strong linkages are apparent are uppermost in the auditor's mind.

Organizations may well focus too much on the EMS manuals and procedures as the main element of the total system. A balance is clearly needed between such a documented system and the improvement techniques and methodology.

It should be kept in mind that the environmental auditors are not looking for a perfect system. They are looking for a workable system – a system that is robust enough to withstand the many other conflicting priorities of a business trying to succeed in today's tough commercial world. Auditors are also looking for a system that is capable of giving performance improvement. During the stage 1 audit, the auditor has a certain amount of flexibility in reporting improvements, development points and advice to the organization; thus the organization understands clearly what is required for the next stage of the certification process (see Chapter 3).

However, the auditor is not allowed to design any part of the management system. This is classed as consultancy and is forbidden by certification body rules and is rigorously monitored by both the certification body and the accreditation body. At the stage 2 audit, the auditor has less flexibility and can only raise opportunities for improvement as a mechanism for system improvement. Having said that, the auditor, even if raising a corrective action request, will ensure that the organization understands what is required to correct the problem. The solution, however, must come from the organization and not the auditor.

Environmental auditing, the law and auditors

It was never the intention of the Standard for auditors to become regulatory authority spies. Certain countries do however make it the law for anyone becoming aware of a breach in regulations to inform the regulatory authority. This places the auditor in a dilemma. He must choose between:

- Compliance with the law

and:

- Respect for the confidentiality agreement that is made with the organization

If an organization is discovered by the auditor to be in breach of regulations, then the auditor will investigate further. Answers will be sought for the following:

- Has the breach been identified by the organization itself? (This would indicate the strength, or otherwise, of the internal audit.)
- Is the occurrence relatively isolated in frequency?
- Have preventive measures been evaluated by the organization?
- Has dialogue taken place between the organization and the regulatory authority?

If the answers to the above are all affirmative, then the auditor can be convinced that there is a commitment to compliance. The regulatory body will not be notified in this case and so the organization's confidentiality is not compromised.

Normally, when an authorization to operate a process is granted by the regulatory authorities, the authorization states, as a condition, that breaches must be notified to the authority. Thus, the auditor should ensure that the organization's system records any infringements of regulations and that the appropriate corrective action is taken.

Consultants

A brief mention is made at this point of the consultants available to assist organizations to achieve ISO 14001. As previously indicated, certification body auditors are limited as to what advice they can give to organizations in order to ensure no conflicts of interest arise during the audit. Therefore, those organizations without the correct resources available internally will need to call on the services of an environmental consultant. There are two distinct types of consultant available:

- Environmental consultants
- Environmental management systems consultants

Each type of consultant is described below.

Environmental consultants

These will probably be an established company, partnership or even an individual, who have been in the business of environmental impact assessment for many years. They will have performed several different types of audit for different reasons (expert witnesses in courts of law during planning applications for construction of roads, buildings or airports, for example) and as such are well versed in all aspects of environmental legislation, land or buildings contamination, impacts on flora and fauna etc. It may well be that over more recent years they have extended their portfolio into performing preparatory environmental reviews for ISO 14001 implementing organizations.

They may not have, as yet, the expertise in assisting companies through the management system element required by ISO 14001 but this is not where their strength lies. Their strength, and where they are best employed, is at the preparatory environmental review stage. If an implementing organization does not have the in-house resources to perform the identification of all environmental aspects required for the preparatory review, then it must seek outside help from such technical experts.

Environmental management systems consultants

Again, these may be an established firm, partnership or individual but they are very likely to be experienced quality assurance consultants who have spread their portfolio into environmental management systems consultancy. They will have many years experience in working with informal management systems and developing them into formalized, documented systems in preparation for certification.

They are very useful to an organization that does not have the resources or the knowledge to drive the system forward following the preparatory environmental review. The consultant's task is to use the preparatory environmental review as a technical specification around which to build an effective management system and to assist the organization to achieve certification.

Summary

The environmental auditor must be an individual with the correct set of skills. This includes strong interpersonal skills, good time management and a confident manner.

Auditing ability is of course paramount and, although qualifications are necessary in this area, the auditor must have a natural inclination to ask questions about a process or situation until satisfied. The qualities of an auditor must go beyond academic excellence and will include skills developed from auditing a wide spectrum of organizations. Technical experts within the team may assist in this broadening of expertise. Being able to identify environmental issues and have a good understanding of the legal framework is also a fundamental requirement. Conflicts of interest may occasionally occur and the ability to make the correct decision without further reference is also of importance.

Auditors are faced with a large amount of information to assimilate during an audit. Some of this information will have been supplied by the organization's personnel. Such personnel may give answers that conflict with other answers from other personnel. The dynamics and culture of the audited organization may not be familiar to them and yet, in a relatively short space of time during the audit, they have to make a judgement as to whether the requirements of ISO 14001 are satisfied.

With the development of other management systems running in parallel with ISO 14001, the idea of the multi-discipline auditor evolving was also suggested.

A brief mention was made about environmental consultants, because some organizations would not be able to obtain certification without expert assistance. A potential organization must recognize that there are two types of consultant and make sure the right type is chosen so that an appropriate system is designed for them.

Appendix I

Glossary

Accreditation

Is a process whereby a certification body is subjected to audit to ensure that it is qualified to issue certificates in specific business sectors based not only upon its auditors expertise, background, training and qualifications but also its internal management processes.

Non-accredited certification bodies can also operate legally. However, they have no higher authority that they are responsible to, such as UKAS in the UK, and in some instances the value of such certification can be debatable.

In the UK when a certificate is issued by an accredited body the certificate bears the 'Crown and Tick'. (See *Crown and Tick* later in this appendix.) This is a logo bearing an image of the royal crown coupled with a large 'tick', and is a respected mark of certification integrity not only in the UK but world-wide.

Accreditation criteria

These are rules and restrictions placed upon certification bodies (see Appendix II).

Assessment process

Is the whole process of audit and certification performed by the third-party certification body. This includes the stage 1 (including the desk-top study) and the stage 2 audits followed by the issue of a report and certificate.

Authorized process

In the UK certain industrial processes are required to be authorized. This is a consent issued either by a local authority or the Environmental Agency.

Emissions need to be monitored and the results of such monitoring are sent to the appropriate authority as well as being kept in the form of environmental records on site available for inspection at any time. Breaches of such authorization must be notified and actions taken to minimize such breaches in the future. The appropriate authority may require certain improvements to be completed by a certain date. Prosecutions can follow in cases of persistent offenders.

Across Europe similar conditions apply although details may be different due to national interpretations of European Directives.

BAT

This is '**B**est **A**vailable **T**echniques'.

Best – meaning that the technique used is effective in achieving high levels of environmental protection.

Available – meaning that the technique has been developed on a scale sufficient to prove that it is economically viable.

Techniques – are methods of operating a process, or designing a plant or are related to using abatement technology.

BATNEEC

A definition arising out of the Environmental Protection Act 1990 – an organization must aim to control emissions using the Best Available Techniques Not Entailing Excessive Costs.

BOD

Effluent entering watercourses may contain bacteria or enzymes for example, which require oxygen for survival. If too much oxygen is used by such bacteria, not enough may be left in the water to sustain aquatic life. Thus discharge consents usually specify maximum amounts of active bacteria, referenced as the biological oxygen demand.

BPEO

A phrase encompassed within the Environmental Protection Act 1990 and is a requirement to minimize pollution by applying the Best Practical Environmental Option.

Brownfield site

A site that is developed for industrial or domestic use that has had previous use. The positive environmental impact is that such use of previously used, and perhaps now derelict land, preserves more virgin or greenfield sites.

Certification

The process of issuing a certificate to an organization that has achieved compliance to a recognized standard and has been independently audited by a third party to such a standard. Such a certificate can then be used as evidence by the organization to demonstrate compliance to any interested party. In the context of this book this is compliance and certification to ISO 14001.

Certification bodies (can be called registrars or registration bodies)

Independent organizations whose sole business is auditing other organizations quality, environmental, health and safety, data protection and other management systems for compliance against national and international standards. Most of the bodies originally assessed to quality standards (ISO 9001:2000 and earlier versions) and have extended their scope into environmental certification and other standards. They have no commercial interests and must be seen to be totally independent so that there can be no question of bias when issuing a certificate.

COD

Effluent entering watercourses may contain chemicals which will use the dissolved oxygen present to attain their oxidized state. If too much oxygen is used by such chemicals, not enough may be left in the water to sustain aquatic life. Thus discharge consents usually specify maximum amounts of active chemicals, via a number referenced as the chemical oxygen demand.

Corrective action requests

During the stage 2 audit, or during routine surveillance visits, by the certification body, if the external auditor discovers areas of concern, or non-conformity, which may jeopardize the integrity of the EMS, a corrective action request will be generated by the auditor. This corrective action request will either be classed as a major or a minor depending upon the severity of the non-conformance. Details may differ from one country to another, from one accreditation body to another and from one certification body to another but as an example for a major corrective action request, a time-scale of 1 to 2 months is given to the client for it to be satisfactorily addressed. This usually necessitates an extra visit by the auditor to observe evidence that corrective action by the client has taken place. Minor corrective actions requests are usually allowed a time-scale of 6 months. They are verified by the auditor by being addressed at the next planned surveillance visit (see also *Non-compliances*).

Contaminated land

This is generally land which has accumulated chemicals, bi-products and waste such that it would present a health hazard to humans or animals. It would either need to be treated 'in situ' or removed to a landfill site. Both options are time-consuming and have high cost implications.

Crown and Tick

In the UK, for historical reasons, the sign of accredited certification used by UKAS is a logo bearing an impression of the Royal Crown plus a 'tick'. There are restrictions placed upon the use of such logos. Thus organizations cannot use the UKAS mark on flags, vans or trucks, primary packaging and promotional goods such as diaries.

Direct aspects/impacts

See *Environmental aspects*.

Discharge consent

This can be considered as a contract between the organization and the local water treatment company whereby, they permit the organization to discharge effluent providing that it meets certain parameters such as minimum or maximum pH, maximum levels of heavy metals, volume and rate of discharge. They may require the organization to monitor and log such discharges and either send in the results periodically, or simply require them to be available for checking by their inspector. If such limits are exceeded, the organization is obliged to inform the authority as to what corrective actions were taken. The authority will possibly charge a fee for treatment of the excess effluent burden, and in persistent or extreme cases will prosecute the organization in a court of law.

EMS (environmental management system)

A management system that enables an organization to control its impacts on the environment. It may well be an informal or fragmented system based on perceived impacts or driven purely by the requirements of legislation. However, the term as understood and discussed in this book means a management system that not only controls its impacts, but takes reasoned and logical steps to minimize environmental impacts and uses the tools of measurement and monitoring. Such a system needs to have a certain minimum level of documentation so that it can be followed by personnel.

Environmental aspects

An element of an organization's activities, products or services that can interact with the environment.

Environmental audit

A management tool comprising a systematic, documented, periodic and objective evaluation of how well an organization's management and equipment are performing, with the aim of contributing to safeguard the environment by:

- Facilitating management control of environmental practices;
- Assessing compliance with organization policies and regulatory requirements.

Although there are several types of environmental audit, suited for different purposes, the above definition related to compliance with ISO 14001 is sufficient.

Environmental impact

Any change to the environment, whether adverse or beneficial, wholly or partially resulting from an organization's activities, products or services.

There are two categories of the above:

1 Direct impacts are those impacts that an organization can directly control;

2 Indirect impacts are those impacts that an organization can only influence by various means. In practice, this means their customers and their suppliers.

'End of Pipe' technology

Defined as rendering the consequences of normal or abnormal pollution less serious by putting technology into place so that the environmental impact is reduced. Best practice is to use resources and technology at the front end of a process to ensure that likelihood of pollution is minimized.

EVABAT

The 'Economically Viable Application of Best Available Technology'. This is another acronym for the use of technology to control environmental impacts without causing an organization financial hardship.

Fauna

Generic name for animal life.

Flora

Generic name for plant life.

Fugitive emissions

The dictionary definition of 'fugitive' can mean 'elusive' or 'moving' or 'roving about', and this definition, within ISO 14001 means fits an uncontrolled, or unexpected release or loss of effluent or more usually, gases into the atmosphere.

Greenfield site

A site that is developed from virgin land. This has many advantages such as being an attractive place in which to work and therefore motivate the workforce. Additionally there may be tax advantages or access to funding for investment. However, it is using up valuable land area and disturbing and destroying wildlife habitats.

Groundwater

Generally means water from beneath the ground's surface. In the context of the protection of groundwater from pollution by dangerous substances, it is taken to be all water that is below the surface of the ground in the saturated zone and in direct contact with the ground or subsoil. Globally, groundwater constitutes 2/3 of the freshwater resource and therefore it is very important not to pollute.

Heavy metals

Refers to the metals, more usually compounds, of lead, cadmium, chromium, vanadium, mercury, nickel. These compounds can exist in solution in waste and have the potential to be taken up by living organisms and tend to accumulate in the bloodstream. They can cause genetic and other health issues.

Indirect environmental aspects/impacts

See *Environmental impact*.

IPC

Is 'Integrated Pollution Control' and relates to the controls exerted by an organization which consider the impacts on the environment as a whole, over all three mediums, air, land and water, i.e. ensuring that solving one environmental problem does not create another elsewhere.

IPPC (Integrated Pollution Prevention Control. EC Directive 96/61/EC (24/10/96))

Is a licence issued by a government body to operate an industrial process. An organization needs to meet the following summarized requirements:

1 All the appropriate preventative measures are taken against pollution;

2 No significant pollution is caused;

3 Waste production is avoided. Where waste is produced, it is recovered. Where this is technically and economically impossible, it is disposed of while avoiding or reducing any impact on the environment;

4 Energy is used efficiently;

5 The necessary measures are taken to prevent accidents and limit their consequences;

6 The necessary measures are taken to avoid any pollution risk once the business activities on the site have ceased.

Landfill site

The location where solid waste is buried underground. There are generally two distinct types:

1 Dilute and disperse sites. Generally of older design which are unlined and allow the free percolation of leachate into the ground.

2 Engineered/dry tombs. Where wastes are encapsulated in a lined void and water excluded.

Some sites produce so much methane from organic matter decomposing, that the gas is used as a fuel for generating electricity on the site.

Landfill tax

In the UK a tax (1996) which is levied upon specified wastes which are disposed of to landfill in an effort to persuade such producers of waste to reduce quantities going to landfill – by being more efficient and recycling.

It is also worth noting that natural landfill sites such as clay and sand pits, disused quarries are becoming scarcer in the UK and newer ones require considerable and costly engineering – which will always result in landfill cost spiralling upwards.

Other European countries have similar taxes to discourage landfill as a solution to waste disposal.

Leachate

As rainfall falls onto landfill sites, it dissolves a whole manner of toxins from the deposited waste, which should not be allowed to enter any groundwater or controlled waters such as streams and rivers. This is collected in ditches or ponds and is either recycled over the site for it to evaporate i.e. to reduce its volume, or is removed by road tankers away to treatment plants elsewhere.

Life cycle analysis

Is based upon a consideration of all the environmental impacts of a product or system from the 'cradle to the grave' i.e. from raw material extraction and processing to manufacture, distribution, usage and ultimate disposal of the product and waste management.

Noise

Generally, environmental complaints about noise relate to its unacceptability at that time i.e. unwanted noise. Thus residents living next to a railway track tend not to complain about the noise of trains, but the intermittent noise of a fork lift truck 'clattering' over bumps perhaps late at night or early in the mornings, will give rise to environmental noise pollution complaints.

Non-compliances

Are discoveries during a management systems audit which demonstrate that a documented procedure or work instruction is not being adhered to by the relevant personnel. Or an objective is not being met through individual targets not being reached. By the process of generating a corrective action request, the 'non-conformance' or 'non-compliance' can be brought to management's attention and following an investigation, steps can be taken to ensure the same problem does not occur again.

No_x

Refers to the several oxides of nitrogen where 'x' is a variable number.

Packaging

Means all products made of any material, of any nature to be used for the containment, protection, handling, delivery and presentation of goods from raw materials to processed goods, from the producer to the user or consumer, including non-returnable items used for the same purpose but only where the products are:

- Sales: primary packaging
- Grouped: secondary packaging
- Transport: tertiary packaging

Packaging directive

This is taken to mean EC Directive 94/62/EC which places obligations upon organizations to meet packaging reduction targets. Collective schemes are in place to relieve smaller companies of this burden – by paying for this service.

Safety management system (SMS)

A management system built up in similar fashion to an environmental management system so that occupational health and safety is managed in a structured way and that continuous improvement in safety practices is achieved.

Significant environmental impacts

A ranking placed upon environment aspects as identified by an organization, as having a greater impact upon the environment than other aspects.

Small to medium enterprises (SMEs)

There is no official definition of what an SME is. However, in Europe when industry statistics are compiled, the definition is based upon employee numbers rather than turnover, profits, market size or number of sites:

Number of employees	Classification of organization
0–9	micro
10–99	small
100–249	medium
250+	large

Therefore organizations with between 10 and 249 employees are SMEs.

However, in the context of this book, the definition of an SME is further refined as follows. An SME :-

- Will have working managers and perhaps owner who have little or no time available for new projects outside of direct business activities;
- Will probably be operating in an informal style and may not have documented management systems;
- Will be deficient in technologically trained personnel i.e. university graduates;
- Cannot afford to employ a single discipline manager such as a quality assurance manager or environmental manager;
- Will undertake 'on the job' training with no structure for identifying training needs.

Therefore, such an organization, although it may be very successful financially, may not have the resources for ISO 14001 implementation.

SO_X

Refers to the several oxides of sulphur where x is a variable.

SSSI

Site of special scientific interest. A wildlife area with special protection offered by British Nature in the UK.

Sustainable development

Many definitions abound but the 'Brundtland Definition' is perhaps the best one. This is named after the Prime Minister of Norway, Gro Harlem Brundtland, who chaired the UN-sponsored World Commission on Environment and Development (WCED) in its report 'Our Common Future' published in 1987 and has its attractions due to its simplicity. Essentially it confirms that continued economic and social development is vital but that this must be without detriment to natural resources (including air, water, land) and bio-diversity (as key resources), especially as continued human activity and further development depend on the quality of these resources. The definition is:

Development which meets the needs of the present without compromising the ability of future generations to meet their own needs.

TC207

An ISO technical committee (TC 207) established to produce internationally recognized and accepted environmental management standards.

TQM

Is 'Total Quality Management'. There are several similar definitions, but essentially TQM can be considered to be the way an organization enacts a philosophy of continually striving to meet its customer requirements and exceeding those expectations. It goes far beyond quality assurance and calls upon enhancement of attitudes of all staff in everything they do.

Sometimes TQM is used incorrectly to describe an integrated management system.

Transfer note (waste transfer note)

Within the UK when waste is passed from one organization to another, the organization accepting the waste must have a written description of it.

It must be signed by representatives of both organizations and contain the following information:

- Description of the waste and quantity;
- The time and date the waste was transferred;
- Where the transfer took place;
- Whether one of the parties is a waste carrier. In the UK and elsewhere, a licence is required from the national equivalent of the Environment Agency to transport waste;
- Whether one of the parties has a waste management licence – again authorized by the equivalent of the Environment Agency.

The above will be embodied in applicable national legislation with reference to due diligence or duty of care.

UKAS (United Kingdom Accreditation Service)

In the UK, this body was set up by the then Department of Trade and Industry (DTI) to provide national accreditation of certification bodies and operates with similar European, American, Australian and Asian bodies. It is an organization limited by guarantee and its members represent various interests in accreditation.

Volatile organic compounds (VOCs)

Covers a wide range of organic chemicals which evaporate at ambient temperatures. They include many commonly used industrial solvents such as white spirit, xylene, toluene, acetone and methyl ethyl ketone (MEK) solvents which are important components of products such as adhesives, varnishes, paints and other coatings.

Waste

Defined as material which is 'discarded'. Material which is capable of reuse as a raw material without need of processing, should not be classified as waste. Waste can cease to be waste after suitable processing to reduce potential harm.

Waste carrier

These are commercial enterprises set up to transport waste safely from the producer to the place of disposal. Generally, they are highly regulated and must have a licence to operate usually from that country's environmental enforcement agency.

Appendix II

Accreditation criteria

Organizations that have their environmental management systems certified to ISO 14001 by an independent body go to the 'Certification Body' or 'Registration Body' or 'Registrar', depending upon the country.

The integrity of certification to ISO 14001 depends to a great extent on the technical abilities and honest working practices of these certification bodies. For this reason, the certification bodies are accredited against requirements drawn up in the international standard ISO IEC Guide 66:1999 'General Requirements for Bodies Operating Assessment and Certification/ Registration of Environmental Management Systems'.

The IAF (International Accreditation Forum) is the international body whose task is to ensure that certification is performed with honesty, competence and consistency from country to country. They do this by facilitation of the national accreditation bodies evaluation of each other, confirming their competence to perform the accreditation function.

In the UK, for example, the accreditation body UKAS assesses the certification bodies on their ability to comply with ISO/IEC Guide 66. Because ISO/IEC 66 is a relatively short document, IAF has developed a guidance document to ensure consistency of interpretation. This is called 'IAF Guidance on the Application of ISO/IEC Guide 66' and was published in 2002 as issue 2 and is quite a lengthy document at 47 A4 size pages.

This appendix focuses on the IAF guidance offered to certification bodies but this guidance can also be useful to organizations about to implement an EMS. From reading this document, implementing organizations can gain some understanding of the way certification bodies are required to operate and some of the reasons for the methodology of preparation for, and performing of, the assessment process. An understanding should also be gained of how accredited certification is highly regulated to ensure that such certification is meaningful and respected.

Only clauses that are relevant to organizations being certified are addressed below, with links shown to chapters of this book.

	Links to chapter in this book	**Clause in IAF guidance**	**Clause in guide 66**
Auditor competence Discusses the certification bodies requirements. To ensure that only appropriately trained and competent auditors are fielded for each assignment.	Chapters 3 & 6	G.4.2.5 to G.4.2.6	Clause 4.2.3.2
Auditor time (i.e. auditor time spent on site) Discusses the process that certification bodies have to employ when quoting the time required for the certification assessment and surveillance visits. Such processes include the use of calculations, charts and graphs to end up with the	Chapter 3	G.4.2.10 plus Annex I	Clauses 5.2.1 5.2.2

	Links to chapter in this book	Clause in IAF guidance	Clause in guide 66
number of man-days required. Criteria include the number of shifts worked, number of employees, nature and gravity of potential environmental aspects, complexity of processes, geographical area to be physically walked by auditor.			
Adding value Discusses how auditors can add value during assessments. For example, by identifying opportunities for improvement as they become evident during the audit without recommending specific solutions.	Chapter 2	G.4.1.23(f)	Clause 4.1.2.0
Auditing methodology Refers to the stage 1 audit – which may only be waived in certain limited circumstances. States that the stage 1 audit acts as a focus for planning the stage 2 audit.	Chapter 3	G.5.3.14 G.5.3.15 G.5.3.17	Clause 5.3
Assessment of the internal audit The extent of the stage 2 audit may be influenced by the degree to which reliance can be placed on the organization's own internal audit.	Chapter 3	G.5.3.20	Clause 5.3
Evaluation of environmental aspects			
States that it is up to the implementing organization	Chapter 2	G.5.3.21	Clause 5.3

	Links to chapter in this book	Clause in IAF guidance	Clause in guide 66
to define criteria for evaluation of significance of environmental impacts. And that it is for the certification body to assess whether the methodology is sound and is being followed.			
Continuous improvement Prompts the organization to define the means by which its policy commitment is achieved and to develop its processes for doing this and measuring progress. It is for the certification body to assess that these processes are sound, and that there are no inconsistencies between policy, objectives, targets and results.	Chapter 2	G.5.3.22	Clause 5.3
Combining with other management systems Suggests that the audit can be combined with audits of other management systems	Chapter 4	G.5.3.24	Clause 5.3
Surveillance and reassessment Refers to sufficiency of ongoing surveillance visits to ensure they are of reasonable frequency. There is reference to frequency being at least once per year. Reassessment is a requirement within ISO/IEC 66 and references maximum period of 3 years.	Chapter 3	G.5.6.1 to G.5.6.12	Clause 5.6

	Links to chapter in this book	Clause in IAF guidance	Clause in guide 66
Multiple sites This gives guidance about the sampling process to be used when auditing multiple sites. Calculations are shown for guidance.	Chapter 3	G.5.3.6	Clause 5.3
Scope of assessment References the interfaces with services or activities which must be addressed if there is a shared site with perhaps common facilities, i.e. common effluent plant.	Chapter 3	G.5.3.3(c) G.5.3.3(d)	Clause 5.3
EMS documentation Guidance is given in that the EMS documentation should make clear the relationship to any other related management system in operation in the organization. It is accepted to combine the documentation for EMS, QMS, SMS as long as the components of the EMS can be clearly identified.	Chapters 2 & 4	G.5.3.2.3	Clause 5.3
Maturity of the system Refers to the expected maturity of the EMS at the time of the assessment. It offers no actual time-scales but expects the certification bodies to derive their own criteria. (Note that in previous guidance documents, 3 months was suggested as the minimum time for maturity.)	Chapter 3	G.4.2.10	Clause 4.2.3.2

	Links to chapter in this book	Clause in IAF guidance	Clause in guide 66
Site This is defined as the main site and references temporary sites (building and construction sites). Or if a location cannot be defined (i.e. for services), the coverage of activities and interfaces.	Chapter 3	G.5.3.5	Clause 5.3

Appendix III

Additional information

Section 1) The ISO 14000 series of standards

All the following standards in the UK can be obtained from:

The British Standards Institute
Customer Services
389 Chiswick High Road
London
W4 4AL

Or contact at website: www.BSI

ISO 14001	*Environmental Management Systems – requirements with guidance for use*
ISO 14004	*'Environmental Management Systems – General Guidelines on Principles, Systems and Supporting Techniques'*

ISO 14004 has been developed to provide additional guidance for organizations on the design, development and maintenance of an EMS. It is not intended to be certified against. It is for those organizations who may feel that they require some additional guidance and background information on the underlying principles and techniques necessary to develop such a system.
These include:

- Internationally accepted principles of environmental management and their application to the development of an environmental system;
- Practical examples of issues arising during the design of the system;
- Practical help sections on system design, development, implementation and maintenance.

ISO 14015:2001 *Environmental Management – Environmental Assessment of Sites and Organizations*

ISO 14020:2000 *Environmental Labels and Declarations – General Principles*
Provides guidelines and principles for self-declaration environmental claims (environmental labelling) made by manufacturers, importers, distributors and retailers of products.

ISO 14021:2001 *Environmental Labels and Declarations – Environmental Labelling – Self-Declared Environmental Claims (Type II Environmental Labelling)*
Guidance on the use of terms for self-declared environmental claims.

ISO 14024:1999 *Environmental Labels and Declarations – Type I Environmental Labelling – Principles and Procedures*
Guidance for establishing a certification programme for third-party environmental claims.

ISO 14025:2000 *Environmental Labels and Declarations – Type III Environmental Declarations – Guiding Principles and Procedures*
Guidance on profiling of product environmental effects.

ISO 14031:1999	*Environmental Management – Environmental Performance Evaluation – Guidelines* Provides guidelines and principles for determining environmental performance of an organization.
ISO 14032:1999	*Environmental Management – Examples of Environmental Performance Evaluation*
ISO 14040:1997	*Environmental Management – Life Cycle Assessment – Principles and Framework* Principles for carrying out and reporting of LCA studies.
ISO 14041:1998	*Environmental Management – Life Cycle Assessment – Goal and Scope Definition and Inventory Analysis* Methodology for definition of goal and scope, performance of LCA, interpretation and reporting.
ISO 14042:2000	*Environmental Management – Life Cycle Assessment – Life Cycle Impact Assessment* Provides guidelines and principles for determining environmental impacts arising from the production, use, and disposal of a product or provision of a service.
ISO 14043:2000	*Environmental Management – Life Cycle Assessment – Life Cycle Interpretation*
ISO 14048:2002	*Environmental Management – Life Cycle Assessment – Data Documentation Format*
ISO 14050:2002	*Environmental Management – Vocabulary* Provides terms and definitions for the above standards.
ISO 14061:1998	*Information to assist forestry organizations in the use of environmental management system standards ISO 14001 and ISO 14004*
ISO 14062:2002	*Environmental Management – integrating environmental aspects into product design and development*

Section 2) Website addresses

www.theacorntrust.org

Acorn Trust. In the UK provides guidance for SMEs on improving their environmental performance.

www.emas.org.uk

EMAS competent body in the UK (The Institute of Environmental Management and Assessment)

www.accreditationforum.com

The International Accreditation Forum website. Access to IAF guidance documents; news; surveys of certification.

www.detr.gov.uk

IPPC guidance notes.

www.ISO14000.com

ISO 14001 information centre.

www.tc207.org

ISO 14001. Work programmes for ISO 14000 series.

www.sentencing-advisory-panel.gov.uk

Legal penalties – gives guidance on the level of fines imposed for breaches of environmental laws such as depositing waste without a licence; polluting controlled waters; illegal abstraction of water; failure to meet packaging, recycling or recovery obligations.

www.sgs.co.uk

SGS United Kingdom Ltd – an accredited certification body.

www.UKAS.com

UKAS – the United Kingdom Accreditation Service.

www.environment-agency.gov.uk/netregs

Designed to assist small businesses to navigate through the maze of environmental legislation.

Includes some sector guidelines for assistance with implementation.

www.hmso.gov.uk
Has links to legislative issues.

www.europa.eu.int/eur-lex
Has links to European Union laws.

Section 3) Regulations and guidance

EMAS

Regulation (EC) No 761/2001 of the European Parliament and of the Council.
This is essential reading for any organization looking for EMAS registration, and is available in the UK from:

> *UK Competent Body*
> *IEMA*
> *Welton House*
> *Limekiln Way*
> *Lincoln*
> *LN2 4US*

The Eco-Management and Audit Scheme: A Practical Implementation Guide
British Library
ISBN 0 946655
Contains the level of detail necessary to implement EMAS.

ISO/IEC Guide 66:1999
General requirements for bodies operating assessment and certification/registration of environmental management systems
The full text of which Appendix II only outlines. In the UK obtainable from:

> *UKAS*
> *Audley House*
> *13 Palace Street*
> *London*
> *SW1E 5HS*

Section 4) Legislation updating

Barbours Health, Safety and Environmental Index
A useful subscription to take out for keeping abreast of legislation.

Barbour Index Plc
New Drift Road
Windsor
Berkshire
BL4 4BR

Croners Environmental Management
Focuses mainly on UK law. Some European and North American legislation is also referenced. Published by:

Croners Publications Ltd
Croner House
London Road
Kingston upon Thames
Surrey
KT2 6SR

ENDS (Environmental Data Services Ltd)
Very topical periodical and published monthly in hard copy and on website.

Finsbury Business Centre
40 Bowling Green Lane
London
EC1R 0NE

NSCA Pollution Handbook
Published annually – UK environmental law.

National Society for Clean Air and Environmental Protection
136 North Street
Brighton
BN1 1RG
ISBN 0 903474 48 4

Appendix IV

EMAS

Introduction

EMAS is the initials of the 'Eco-Management Audit Scheme' which is a European environmental management systems standard. It is essentially European Council Regulation (EEC) No. 1836/93 'allowing voluntary participation by companies in the industrial sector in a Community Eco-Management and Audit Scheme'. It was published in its entirety in Official Journal L168 dated 10 July 1993. It was revised and republished as regulation (EC) No. 761/2001 of the European Parliament and of the Council, dated 19 March 2001.

This chapter is primarily intended for those organizations who wish to implement EMAS but may be unsure of the similarities and differences between EMAS and ISO 14001. Are two discrete systems required? Are two systems of documentation required? Does one system complement the other? These and other questions will be answered in this chapter but it must be understood that this chapter is included for breadth of information only. This book does not offer extensive or detailed implementation advice. However, the differences and similarities between the two environmental Standards are highlighted. References to further reading and information sources will be found in Appendix III.

It must also be said that if an organization has understood the concepts of ISO 14001, then implementation of EMAS should present no difficulties or problems as the same fundamental methodology and terminology is used by both standards. The approach taken in this chapter is that of assuming that an organization has already achieved ISO 14001, and now wishes to obtain EMAS. This approach is certainly the prevalent one in all the organizations that are registered to EMAS to date.

History, concepts and reasons for registration to EMAS

The idea formulated for the development of EMAS as a recognized European standard for environmental management was that an organization that decided to register could say very publicly that they had nothing to hide regarding environmental issues. They were stating that they did have impacts on the environment but were taking positive steps to reduce such impacts. A great deal of information as to how they were doing this would be publicly available.

This should include verifiable numerical data related to raw material and energy usage, by-products' and waste products' volumes and descriptions. Certainly the concept of publishing such information pro-actively was thought to be preferable than allowing bad publicity to distort the real situation.

In this way, the organization has the opportunity to put its case first and demonstrate, in a very public way, its achievements in environmental issues. Inherent in the drafting process of EMAS was the hope by the European Council that the publication of such detailed information would induce companies not just to achieve compliance with the law but to go well beyond.

As referenced above, EMAS had its beginnings in 1993 and was revised and republished in 2001, with many of the restrictions removed such as being no longer restricted to certain industrial sectors and companies with multiple sites being accommodated. (Clause 7 of the regulation stating 'EMAS should be made available to all organizations having environmental impacts, providing a means for them to manage these impacts and to improve their overall environmental performance.')

Implementation of EMAS – links to ISO 14001

As indicated in the introduction, only broad guidance is offered by indicating the main steps in the process of implementation. If an organization seeks to implement EMAS, further reading and detailed steps are available from the national Competent Authority for each country within Europe.

The following steps should be considered for an organization starting from the beginning to achieve EMAS:

1 Conduct an environmental review of all activities. (This is akin to the PER suggested for ISO 14001 implementation.) Annex VII of the regulation details this.

   However, the regulation does recognize that if the implementing organization has already achieved ISO 14001, then this requirement is probably not required.

2 Carry out environmental auditing in accordance with the requirements set out in Annex II of the regulation.

   This will not be unfamiliar to organizations already certified to ISO 14001 and who therefore have carried out a programme of internal audits. However, there is one item relating to the audit cycle being completed within a timeframe – 'at intervals no longer than 3 years'.

3 Prepare an environmental statement following the requirements of Annex III of the regulation.

4 Have the environmental statement, management system, audit procedure validated by the environmental verifier to ensure it meets the requirements of Annex III of the regulation.

   ISO 14001 now forms the basis of the management system requirements.

5 Forward the validated environmental statement to the competent body of the Member State in which the organization seeking registration is located and after registration, make it publicly available.

Note that in order to maintain registration (similar to the continuous assessment or surveillance requirements of ISO 14001) the organization must forward yearly validated updates of its environmental statement to the competent body and make them publicly available.

Because this book has addressed in previous chapters the first two steps above, the environmental statement requirements are expanded upon in the next section.

Environmental statements – structure

EMAS is all about organizations taking responsibility for their impacts upon the environment, being committed to continual improvement in environmental performance, and reporting that performance publicly. The environmental statement is the means by which an organization can communicate publicly its progress in managing and improving the environmental impacts of the site's activities.

To register under EMAS, the organization must establish an appropriate mechanism for reporting its progress and achievements.

In essence, the environmental statement provides the organization with the opportunity to produce a balanced account of the overall aims, objectives, targets and other influences affecting the site's overall performance. The public statement will be held with the national regulatory bodies and will be available locally to anyone who wishes to view it. The regulation states that organizations are encouraged to use all methods available for publicity such as electronic methods, the Internet, publications, libraries etc.

The information within the statement needs to be understandable. It must set out the nature of the business, the environmental performance it is trying to achieve and the progress it is making. The content of the statement is specified in the EMAS regulation minimum requirements. Just as in ISO 14001 the key word and concept throughout was 'significance', the key word here is 'balance'. Providing the public with information about the site's environmental performance should not be seen as only confessing to failures or only focusing on success.

The format of the statement does not have to be of any specified length or presentation style. EMAS does require that the statement should be able to be read and understood by the public. This implies the minimum use of 'jargon' and technical terms which may be hard for a member of the public to comprehend. Any technical material might be included in an attached appendix. One registered publishing organization produces their statement on glossy art paper illustrated with photographs. This is appropriate and practical for them to produce. A smaller organization with limited resources would be wiser to opt for a simpler approach. However, the technical content must be of a certain minimum standard as outlined below.

The requirement for the public to understand such technical content may, at first sight, represent a significant challenge for the organization. The technical processes of the organization may be very complex and may not be easily described in terms familiar to all members of the public. The public concerned represents a wide cross-section of society, with equally wide levels of understanding of technical issues.

Although the stakeholders will have an interest in the contents of the environmental statement, the 'public' for which the statement is also intended needs some explanation. The term 'public' will be different for larger and multinational organizations compared to smaller, locally based organizations. The target public for a smaller company may well be only the local public – passers by, local interest groups etc.

Larger organizations with a high public profile may have to cater for the more sophisticated members of the public. Such readers of the statement will require a reasonable amount of technical detail to gain an in-depth understanding of the environmental issues. An environmental statement should contain data and information that is reliable, for it will be subject to verification. The statement should also cover all the significant environmental issues of relevance to the site. It is also necessary to define the basis on which issues are omitted from the statement. If estimates are used then the process of estimation should be described and shown to be technically valid.

The statement itself should be prepared and made available as soon as the environmental management system is in place and the audit programme has commenced. Subsequent statements should be produced on completion of each audit cycle.

The regulations specify minimum requirements (Annex III of the Regulation) and should typically follow a standardized format as follows:

a) *A clear and unambiguous description of the organization registering under EMAS and a summary of its activities, products and services and its relationship to any parent organizations as appropriate.*

   This description of the organization's activities should be understandable with a minimum of technical terms and industry jargon.

b) *The environmental policy and a brief description of the environmental management system of the organization.*

   This should explain what the management at the site is trying to achieve, and how. This should be both in general terms (for example, the policy aims and objectives) and in specific terms (the programme of actions, targets and deadlines). The policy, programme, objectives and targets should relate to and address the significant aspects identified below.

c) *A description of all the significant direct and indirect environmental aspects which result in significant environmental impacts of the organization and an explanation of the nature of the impacts as related to these aspects.*

   As well as identifying and describing the relevant environmental issues in non-technical language, there needs to be an accompanying explanation as to how and why these were assessed as being significant. Annex VI of the regulation prescribes those aspects which need to be considered as well as offering guidance on the criteria for assessing significance.

d) *A description of the environmental objectives and targets in relation to the significant environmental aspects and impacts.*

   It is worth noting that Annex IB, lays out some specific issues which need addressing for EMAS, the most significant departure from ISO 14001 being a focus on the improvements in actual environmental performance and this will need to be reflected within the data in the statement.

e) *A summary of the data available on the performance of the organization against its environmental objectives and targets with respect to its significant environmental impacts. The data should allow for year-by-year comparison to assess the development of the environmental performance of the organization.*

This is not intended to be a mere list of pollutant emissions and quantities of materials used but rather a summary presented in a way which can be easily read, interpreted and conclusions easily reached. Grouping emissions into their environmental impact is one way. Any trends to date should be indicated.

Organizations may use relevant existing performance indicators making sure that the indicators chosen:

1 give an accurate appraisal of the organization's performance;

2 are understandable and unambiguous;

3 allow for year on year comparison to assess the development of the environmental performance of the organization;

4 allow for comparison with sector, national or regional benchmarks as appropriate;

5 allow for comparison with regulatory requirements as appropriate.

For example, performance figures should be normalized to production output. This is to negate the distortion created if, for example, a company could demonstrate that waste fell by 50% in a period of 12 months. This would be a fine achievement but meaningless if it had been achieved by halving production output. Thus a balanced set of figures must show details of toxic and hazardous waste, use of utilities, any recycling efforts, emissions to air, discharges to sewer, raw materials usage etc.

f) *The name and accreditation number of the environmental verifier and the date of validation.*

The name of the accredited verifier must be included on the statement submitted to the Competent Body. Copies of the statements intended to be made publicly available must also include this name. The verifier is the certification body who confirms that the content of the statement is a true record of the state of affairs on the site, and evaluates the performance of the environmental protection system.

ISO 14001 certification bodies can also act as verifiers providing that they conform to EMAS-specific accreditation criteria. Individual EMAS auditors are called verifiers. The audit work they perform is very similar to the work of ISO 14001 auditors. However, they are required to have much more in-depth knowledge of environmental legislation and a minimum amount of experience in the environmental field. Such experience covers a broad range of environmental issues.

EMAS or ISO 14001?

An organization has three choices when it comes to deciding whether to go ahead and implement an environmental management system. It can choose one of the following:

1. Obtain ISO 14001 initially in its own right and then, after a suitable period of time (dictated by the organization itself), register for EMAS.

   The fact that the organization has an existing environmental management system is taken fully into account during the EMAS verification to the point that additional verification of the system is not performed (that is, part of the requirements of EMAS have been met). The verification will then focus very much on validation of the environmental statement.

2. Be assessed to ISO 14001 and verified to EMAS at the same time. EMAS verification and ISO 14001 assessment can be performed concurrently by a certification body using a team which includes an EMAS verifier.

   Failure to reach some of the requirements of EMAS (in the environmental statement) may mean that certification to ISO 14001 is allowed but there is a delay in obtaining EMAS until corrective actions have been taken.

   Failure to meet the requirements of ISO 14001 will delay both certifications as all the elements of ISO 14001 need to be in place to achieve EMAS.

3. Register only to EMAS. The verification team will of course look at all the elements required of an environmental management system (that is, ISO 14001) but should the audit be successful, only EMAS registration is granted.

Obviously, an organization achieving both standards not only has to produce the requisite environmental statement at the correct times but is also subject to ongoing annual or 6-monthly surveillance/continuous assessment visits in order to comply with the requirements of ISO 14001.

When successfully verified by the external verifier, the company will be awarded the ECO Audit Logo which must always be accompanied by a schedule detailing the sites involved. Several alternative phrases and templates are used (as appropriate) and these are described in the Regulation itself.

Summary

EMAS is a European standard and only time will tell whether it will have the same recognition, acceptability and credibility outside Europe as ISO 14001. Both standards have the same underlying principles but EMAS goes a step further in requiring the organization to publish data on its environmental performance. This was seen by the writers of EMAS to be a desirable approach for organizations to demonstrate that they were environmentally responsible and had nothing to hide. It was hoped that the uptake, on a voluntary basis, would be high. Some organizations, however, saw EMAS as a threat – reasoning that publishing such data and thereby revealing commercially sensitive information about their processes might give their competitors an unfair advantage.

Index

Page numbers in bold indicate a more detailed reference for that entry.